Reviews of Physiology, Biochemistry and Pharmacology 141

Springer-Verlag Berlin Heidelberg GmbH

Reviews of

141 Physiology Biochemistry and Pharmacology

Special Issue on Water Transport Across Biological Membranes
Edited by R. Greger and W. Rosenthal (Guest Editor)

With 34 Figures and 12 Tables

 Springer

ISSN 0303-4240

ISBN 978-3-662-31206-3   ISBN 978-3-540-48082-2 (eBook)

DOI 10.1007/978-3-540-48082-2

Library of Congress-Catalog-Card Number 74-3674

Production: PRO EDIT GmbH, 69126 Heidelberg, Germany
Printed on acid-free paper – SPIN: 10718045   27/3136wg-5 4 3 2 1 0

Contents

Indexed in Current Contents

Molecular Biology of Aquaporins

Kenichi Ishibashi[1], Michio Kuwahara[2], and Sei Sasaki[2]

[1]Department of Pharmacology, Jichi Medical School, Minamikawachi,
Kawachi, Tochigi 329–0498 Japan; and
[2]Internal Medicine II, Tokyo Medical and Dental University

Contents

Water transport across the plasma membrane is important for fluid absorption and/or secretion as well as cell volume regulation. Water is transported through biological membrane in two mechanisms: diffusional transport and water channel mediated transport. Despite its low lipid permeability, water has high permeability for almost all cell membranes (a possible exception is the apical membrane of thick ascending limb of Henle of kidney) to prevent the development of osmolarity gradient across cell membranes. This water flux occurs by diffusion driven by osmotic and hydraulic forces that arise from active transport of ions and osmotic solutes. On the other hand, highly specialized cells that transport bulk of water rapidly are equipped with aqueous transmembrane pathways, pores. They are named water channels or aquaporins (AQPs). Recent molecular clonings of water channels have given the insight into molecular mechanisms of water transport in highly water permeable membranes. Many isoforms of AQPs are identified and they distribute widely. AQPs belong to MIP family. MIP family was named after the first sequenced member of this family, the major intrinsic protein (MIP) of lens fiber cell membrane. The number of the members of MIP family is now increasing mainly due to EST and genome projects. Every organism seems to have some members of MIP family with limited exceptions. Some of them also function as small non-ionic solute channels.

In this review, we will focus on the diversity of aquaporins as members of MIP family (AQP/MIP family). More extensive current reviews on the functional aspects of AQPs are presented in this journal and in the literature [1–6]. We will update the current members of AQPs in mammals. Much more expanding members of plant AQPs (at least 25 in Arabidopsis thaliana)[7] and AQPs in lower organisms such as C. elegans, yeast, and bacteria will also be included. As some of the genome projects have been completed and all members of MIP family in these organisms are identified, it will be possible to discuss the evolutionary aspects of MIP family. The references cited in this review are limited only to the recent ones. The original and older references are easily found in the recent related references and the previous extensive reviews in this field [1, 7].

1
The Characteristics of AQP/MIP Family

Several characteristic features of AQP/MIP family proteins were noticed when the name of MIP family was proposed [8]. However, these original features may need to be revised after the accumulation of many members by PCR-based cloning and the genome projects. The features will be updated to incorporate the deviated members of this family. The ubiquitous distribution of MIP family suggests its ancient origin. Several variants of this family are expected to have come out in the course of its evolution. They can be subgrouped to highlight the phylogenetic feature of MIP family. The subgrouping may also reflect the functional diversity of MIP family.

1.1
Relatively Small Proteins

Most members of this family are less than 300 residues in length, usually 250–280. This characteristic of small size has been helpful to clone, analyze, and characterize the members of this family. However, a few eukaryotic members are substantially longer. Two members of the yeast have 646 and 669 amino acids, and one member from an insect has 700 amino acids. The extra amino acids of these proteins are hydrophilic segments and located in cytoplasm. The longer version of this family has not been identified in bacteria, plants, and vertebrates. As the genome project of C. elegans has been completed, the absence of longer members in C. elegans is evident. It is possible that such a long version has been lost in the course of evolution. It may be possible that the extra hydrophilic segments may have become separate proteins and can be found in the genome. It is tempting to speculate that these proteins become associate proteins with MIP family to modulate their functions.

1.2
Conserved NPA Motif

At the time of MIP family identification, the sequence comparison among members revealed the two regions highly conserved. These regions were highlighted by three amino acid residues (Asn-Pro-Ala) with surrounding several amino acids. These relatively hydrophobic segments are named NPA boxes. As a result of tandem repeats (see below), two NPA boxes are present in each protein. The first NPA box is usually NPAVT and the second NPA box is NPARS/D. The first NPA box is often preceded by five-residue sequence ISGAH with a single variable hydrophobic residue in between. Following the analysis of all members of MIP family then available, a signature sequence of MIP family was proposed [8]; (HQ)-(LIVMF)-N-P-(AST)-(LIVMF)-T-(LIVMF)-(GA) for the first NPA box because the first NPA box is more conserved among MIP family. However, the analysis of the current database of MIP family reveals much more variation of NPA box motif. Examples of the several variants of NPA include NPS, NAA, NPI, NPV, NCA,

K. Ishibashi et al.

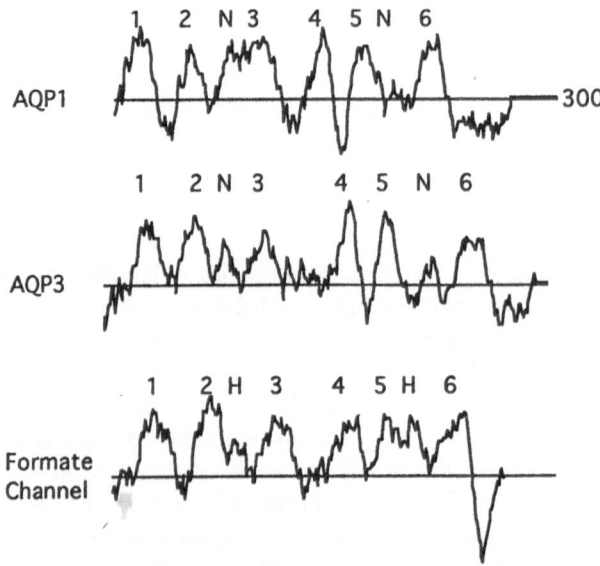

Fig. 1. Hydropathy profiles of two AQP/MIP family (AQP1 and AQP3) and a putative formate channel from Bacillus subtilis by Kyte-Doolittle hydrophobicity analysis using a window of 13 residues. Y-axis shows the hydrophobicity index. X-axis shows the number of amino acids from 1 to 300. The numbers are putative transmembrane segments. N represents NPA box, while H represents small hydrophobic segments that may correspond to NPA box of MIP family. Note that the formate channel is not a member of MIP family but has a similar hydropathy profile. AQP1 belongs to CHIP group whereas AQP3 belongs to GLP group. Note that the distances between transmembrane 3/4 and between transmembrane 5/6 of AQP3 are longer than those of AQP1

NLA, DPL, NPT, NPC, NPG, and SPL. The above signature of the NPA box will be expanded as (HQAMFS)-(LIVMFTCS)-(NCDS)-(PCLA)-(ASTLCIVG)-(LIVMFPGA)-(TSIPV)-(LIVMFW)-(GALTSDYFME). However, such expanded signature sequence will identify proteins irrelevant to MIP family because the criteria for MIP family depend on the overall sequence similarity and not simply on the signature sequence. Therefore, it is necessary to verify by other features of MIP family (see below) to call them members of MIP family. Until now, it has been easy to identify the members of MIP family because they have the typical signature sequences of MIP family. It should be realized, however, that the boundary between MIP family and other six-transmembrane proteins is becoming less clear because of the identification of new proteins that have low homology with the current members of MIP family [Ishibashi et al. unpublished observation]. Such new proteins will benefit us in contemplating structure-function relationship of

MIP family. Furthermore, some membrane proteins that have no sequence homology with MIP family may have similar functions as MIP family. For example, the formate channel of bacteria has a very similar hydropathy profile with MIP family. It has six transmembrane domains with small hydrophobic segments in loop B and E, which may correspond to NPA boxes although the primary sequences are not related (Fig. 1). It will form a pore for small solute such as formate. In the near future, MIP family will be included in a small solute channel superfamily. The data from three-dimensional crystallography will bring the seemingly unrelated proteins from primary sequence together. Some of them may be water channels. Such novel water channels will reveal water channel specific segments when compared with current AQPs. Such water channels can be identified in the cells whose water permeability is high but no AQPs are identified. One of these organs is skeletal muscle, which functions as a water reservoir and whose plasma membrane has high water permeability. Only AQP4 has been identified in skeletal muscle but its expression is sporadic [9]. The muscle fibers that do not express AQP4 may have novel water channels.

1.3
Tandem Sequence Repeats

The first half of amino acid residues of MIP family shares a considerable similarity to that of the second half (~23%), suggesting that the intragenic duplication produced MIP family proteins. This sequence similarity is lower than the similarity between the members of MIP family per se. Within a member of MIP family, only the residues around NPA boxes are conserved. Therefore, the intragenic duplication occurred before the divergence of MIP family. The first halves of MIP family are more conserved than the second halves among the members of MIP family. Such variations of the second halves may be responsible for the functional divergence. An insert of amino acid residues is typically found in the second halves in glycerol-permeable aquaporins (see below). As the halves of MIP family consist of three membrane spanning regions, the intragenic duplication produced the integral membrane protein with six membrane spanning domains (see below). The tandem repeats of the odd numbered transmembrane domains bring the NPA boxes to the opposite side of the membrane to interact with each other as proposed in hourglass model and provide the structural basis for the absence of rectification of its transport. Another insert of amino acid residues is found between this junction in glycerol-permeable aquaporins (see below). The repeats of NPA boxes made it easier to clone new members of this family by amplifying DNA sequences between these NPA boxes whose

6   K. Ishibashi et al.

sequences are used for constructing a set of PCR primers. However, as there are one or two introns between NPA boxes and these are usually long, it is not easy if not impossible to clone MIP family from genomic DNAs. Furthermore, as the conservation of NPA boxes are not so strict, we may miss some members of this family if we depend solely on this PCR method. In the era of genome projects the database search by BLAST program is a more efficient and widely applicable method.

1.4
Six Membrane Spanning Domains

The hydrophobicity analysis of the amino acid sequence of MIP family revealed that they are highly hydrophobic and appear to span the membrane six times as α-helix. The absence of a signal peptide at N-terminus suggests that both N- and C-termini are present in cytoplasm. The hydrophobicity plots of two representative AQP/MIP members are shown in Fig. 1, and the general topology of this family is presented in Fig. 2. The transmembrane domains are named from N-end successively I, II, III, ---, and the loops

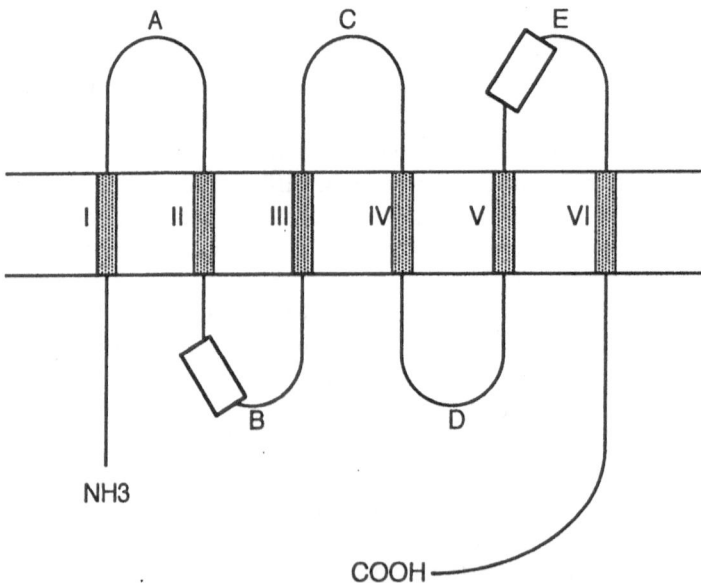

Fig. 2. Menbrane topology of AQP/MIP family. The transmembrane domains are named from N-end successively I, II, III, ---, and the loops connecting these transmembrane domains are named A, B, C, ---. The NPA boxes shown by blank boxes are located at loop B and E

connecting these transmembrane domains are named A, B, C, ---. The two internal repeats of three transmembrane segments with the opposite orientation produce a six membrane spanning protein. The NPA boxes are located at loop B and E. The validity of this topology was confirmed by introducing the mutation susceptible to enzyme digestion and segment-specific antibodies. The cytosolic C-terminus is usually longer than the N-terminus and the both may have phosphorylation sites by protein kinases. The loop D is short but highly hydrophilic in many members. Such a hydrophilic segment is not present in the corresponding loop A of the first halves.

Hydropathy profile of a formate channel of bacteria was compared with MIP family members (Fig. 1). It has similar hydrophobic profile to that of MIP family. Two small hydrophobic segments (indicated by H) are present in loop B and loop E, which seem to correspond to NPA boxes of MIP family. This hydropathy profile similarity between the formate channel and MIP family suggests that the basic structure of the formate channel is similar to that of MIP family. It will be interesting to examine whether MIP family functions as a formate channel and/or a formate channel functions as a water channel.

2
The Members of AQP/MIP Family

A recent extensive review of 84 MIP family members revealed that they fall into twelve subfamilies [8]. They and others including ourselves noticed that MIP family could be divided more generally into two groups based on the primary sequences [6, 8, 10–12]. We name the first group CHIP group representing the first member of water channel (CHIP28/AQP1 of mammals), and the second group GLP representing the first member of glycerin transporter (GlpF of E. coli). Functionally, this division seems to represent the distinct groups. The members of CHIP group are water-selective water channels, whereas the members of GLP group are glycerol permeable water channels. However, the sequence basis for the functional difference has not been established and some exceptions are noted. For example, nodulin 26 is included in CHIP group from its primary sequence, but it transports water and glycerol [13]. Therefore, our grouping of MIP family is solely on the basis of the primary sequences and the naming relating to the function may not be appropriate at this stage. Interestingly, Gram-negative bacteria have a set of MIP family proteins representing each subfamily members. It is tempting to speculate that current diversity of MIP family is derived from the gene duplication including genome duplication from two bacterial ancestor proteins. When compared, two groups have conserved residues around the

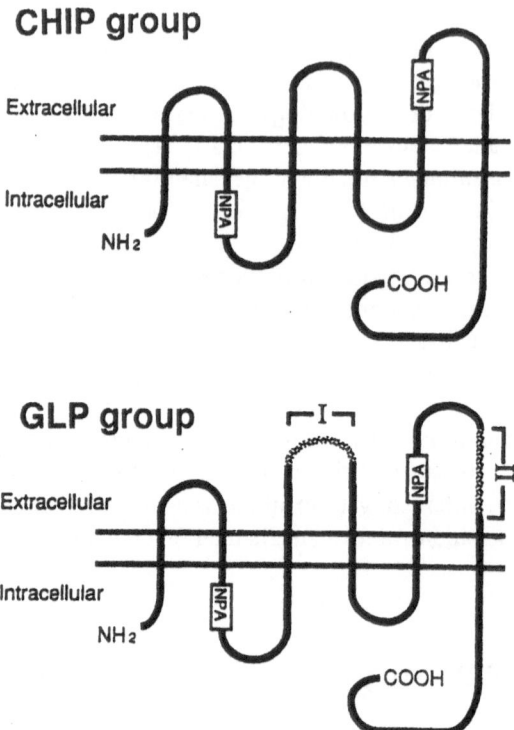

Fig. 3. Menbrane topology of two groups of AQP/MIP family; CHIP group and GLP group. GLP group has extra amino acid residues (~15 amino acids) at loop C and loop E

second NPA box relatively specific to each subgroup. CHIP group has N-P-A-(RVL)-(SA), while GLP group has N-P-A-R-D. The division is also possible when the overall structures are compared (Fig. 3). GLP group has two inserting residues at loop C and loop E. These features are convenient for division, but definitive subgrouping should be based on the overall homology with representative members of each group. Most of MIP family members can easily be divided into these two groups. However, the common ancestor for both groups may be present. Furthermore, the hybrid of both groups may also be present. Currently, these members have not been identified. Such possibility should be kept in mind in screening AQP/MIP family in genomes of organisms.

Table 1 summarize the members of MIP family divided into two subgroups. They will be considered on the basis of organisms below.

Table 1. Distribution of Two Groups of AQP/MIP Family

Organisms		CHIP	GLP
Archaea	Archaeoglobus fulgidus[a]	1	0
	Methanobacterium thermoautotrophicum*	1	0
	Methanococcus jannaschii[a]	0	0
	Pyrococcus horikoshii[a]	0	0
Eubacteria	Escherichia coli[a]	1	1
	Haemophilus influenzae[a]	1	1
	Shigella flexneii	1	1
	Bacillus subtilis[a]	0	1
	Mycoplasma genitalium[a]	0	1
	Borrelia burgdorferi[a]	0	1
	Synechocystis[a]	1	0
	Treponema pallidum[a]	0	0
	Mycobacterium tuberculosis[a]	0	0
Fungus	Saccharomyces cerevisiae[a]	2	2
Plant	Arabidopsis thaliana	25	0
Nematode	Caenorhabditis elegans[a]	3	5
Insect	Drosphila melanogaster	7	0
Vertebrates	Zebrafish	5	3
	Frogs	1	1
	Human	7	3

[a] The genome project has been completed.

2.1
Bacteria

Bacteria are divided into two groups; archaea and eubacteria. Several genome projects of bacteria have been finished. Such bacteria are considered here. In some archaea, Methanococcus jannaschii and Pyrococcus horikoshii, no MIP family protein has been identified. However, some other archaea, Archaeoglobus fulgidus and Methanobacterium thermoautotrophicum, have a member of MIP family belonging to CHIP group. The presence of GLP group in archaea has not yet been documented. Whether GLP group was lost in archaea or CHIP group is older than GLP group, remains to be clarified. Since some bacteria do not have any MIP family protein, MIP family is not indispensable for the life of bacteria. In eubacteria, two MIP family members are usually present. In E. coli, AQP-Z, a CHIP group and GlpF, a GLP group are present. No other MIP family proteins are present in E. coli. Another closely related Gram-negative bacteria, Haemophilus influenzae and Shigella flexneii, also have two MIP family proteins; a CHIP group and a GLP group. On the other hand, Bacillus subtilis, Mycoplasma genitalium, and Borrelia burgdorferi have a single MIP family protein that belongs to GLP group. GLP group proteins in bacteria are present within an operon that also encodes proteins concerned with glycerol catabolism. However, CHIP group proteins are not associated with any operons. Therefore, it is possible that CHIP group proteins may have been acquired by horizontal transfer. A cyanobacterium, Synechocystis also has a single member MIP family but it belongs to CHIP group. Treponema pallidum and Mycobacterium tuberculosis have no MIP family protein. Again, MIP family is not always present in Bacteria. Only gram-negative bacteria seem to have a set of MIP family. Therefore, MIP family may not be important for the survival of eubacteria as well. The null mutants of GlpF have only a problem with slow growth in glycerol-rich medium and the null mutants of AQP-Z exhibit reduced growth rate leading to a smaller colony size [14]. Therefore, MIP family proteins can modulate the life of bacteria but they are not critically important for survival. Alternatively, some of MIP family proteins in bacteria may have been changed so much that the current BLAST search can not identify them as MIP family. Some "hypothetical membrane proteins" in database may have replaced the function of MIP family in these bacteria that have no obvious MIP family proteins.

2.2
Yeast

The genome project of Saccharomyces cerevisiae has been completed. The analysis of its database revealed four members of MIP family. Two of them belong to GLP group and the other two belong to CHIP group. One of CHIP group has frame shift mutation and may be a pseudogene. Both GLP members have unusual long N- and C- termini. However, these extra residues are not sequence-related with each other and no homologous residues are found in database. These extra residues may have been attached to these two GLP members after its gene duplication in yeast. One of CHIP group (AQPY1) was found to have been mutated in laboratory yeast strains. No water channel activity is detected in the vesicles prepared from this strain, but the growth of this strain seems normal [15]. Therefore, CHIP group in yeast may not be indispensable. On the other hand, the loss of a member of GLP group (FPS1) causes the defect of cell fusion leading to the defect of mating possibly due to the osmotic unbalance [16].

2.3
Plants

In plants, all MIP family proteins belong to CHIP group and no members of GLP group have been identified yet. As the genome project of a model plant, Arabidopsis thaliana, is still in progress, plants may have GLP members that are not yet identified. The absence of GLP members in plants is strange because the yeast has two members of GLP group. The unusually long GLP members in yeast may have some relevance for the absence of GLP group in plants. The number of CHIP group in plants is extremely large. It was claimed that independent 23 members are identified in Arabidopsis thaliana by the analysis of EST clones [17]. However, some of them were redundant or erroneously classified as new genes because of sequence errors. A recent our survey in non-redundant protein database of Arabidopsis thaliana (13,299 sequences) indicated that there are 25 MIP family members. All belong to the CHIP group. These CHIP members can be divided into three subfamilies, plasma membrane type (PIP), tonoplast type (TIP), and sym-biosome type (NOD). There are ten members in PIP subfamily, nine members in TIP subfamily, and four members in NOD subfamily. Although the distribution of NOD is not restricted to nitrogen fixing legumes, the limited diversity of NOD is noteworthy. In the primary sequence, PIP and TIP are more similar with each other than with NOD. As the genome structure of TIP is simpler (usually a single exon-intron boundary) than PIP, TIP may be more ancient than PIP. As nodulin 26 also transport glycerol [13], the function of GLP group may be overtaken by NOD subfamily. The absence of GLP group in plants may be explained by the emergence of a new subgroup in CHIP group, i.e. NOD subfamily in plants. Furthermore, some of PIP and TIP members are shown to transport glycerol as well as water [18, 19]. Therefore, the dichotomy of MIP family seems to have lost its functional basis. More subtle difference in the sequence and/or structure may cause the difference of selectivity; water vs. glycerol.

2.4
Invertebrates

The genome project of C. elegans has been completed. The conventional MIP signature sequence identified eight MIP family members. Five of them belong to GLP group and the three seem to belong to CHIP group. However, this division is not clear because CHIP group members of C. elegans are divergent from other CHIP group members. One of the GLP group members has been functionally characterized. It (AQP-CE1) stimulated water and urea transport. However, it failed to stimulate the glycerol transport [20]. This transport characteristic is unusual in MIP family. Such transport character has also been reported in mouse AQP8 [21] and human AQP9 [22]. The pore size heterogeneity may have produced such a selectivity difference (see below). As the mammals have currently three members of GLP group, five members of GLP group in nematode is noteworthy. Mammals may have more GLP members that are currently not identified.

In insects, several MIP family proteins are identified. They all belong to CHIP group. However, as the search for MIP family in insects is currently preliminary, GLP member will be found in the near future when the genome project of Drosophila has been completed. Recent our survey in database of Drosophila (97,564 sequences) indicated the presence of seven members of MIP family that all belong to CHIP group. Bib (Big Brain) of Drosophila has long C-terminus hydrophilic segment. Bib is the only CHIP member with long extra residues. The predominance of CHIP group in insects is similar to that in plants.

2.5
Vertebrates

In the database of the genome project of Zebrafish, we identified eight members of MIP family. Five of them belong to CHIP group and three of them belong to GLP group. In amphibian, AQPs of toad bladder are well characterized and some members are identified. Since vasopressin-regulated water permeability was discovered 40 years ago, the toad bladder has become the principal model for studies of cell biology of vasopressin-regulated water transport. FA-CHIP was first identified and later AQP-TB (AQP-t1) was identified from frog urinary bladder. Both (just a species difference) are ortholog of mammalian AQP1 and expressed at the endothelium of vessels including urinary bladder [23]. Interestingly, they are not expressed at frog red blood cells (AQP1 is rich in mammalian red blood cells). The transcript of FA-CHIP increased with salt acclimation but that of AQP-TB did not increase with dehydration or vasopressin treatment. It is possible that toad bladder may have a real ortholog of AQP2, namely vasopressin-regulated AQP that is not yet identified. The ortholog of AQP3 (78% identity), a GLP member, was recently cloned from Xenopus kidney [24]. The expression of AQP3 homolog in oocyte explains the endogenous glycerol transport of oocytes. The presence of MIP family in reptiles and in birds has not yet been examined in the literature.

AQP/MIP family has been extensively examined in mammals. Ten members have been identified in human and rat. They can be divided into two groups. They are aligned in Fig. 4. AQP0, 1, 2, 4, 5, 6, 8 belong to CHIP group, and AQP3, 7, 9 belong to GLP group. Among CHIP group, AQP8 is less similar to other CHIP group members. Furthermore, the exon-intron boundaries of AQP8 are also different from those of other mammalian CHIP members [25]. The primary sequence of AQP8 is more similar to that of that of TIP family in plants. Therefore, AQP8 may be a separate subfamily in CHIP group in mammals. The genome structures of other CHIP proteins are similar with 4 exons while those of GLP group are similar with 6 exons. The chromosomal mapping revealed that AQP0, 2, 5, 6 colocalized at the same chromosome locus (12q13) and that AQP3 and 7 at 9p13. These AQPs may have been formed by local gene duplication. These characteristics and tissue distribution of mammalian AQPs are summarized in Table 2 and Table 3. Some of the characteristics of each mammalian AQP will be discussed one by one.

AQP0 is the first member of MIP family to be cloned. AQP0 is exclusively expressed in the fiber cell membranes of eye lens, which constitutes ~60% of the integral membrane protein of the lens. Thus it was originally named MIP26 (major intrinsic protein of 26 kD). Compared with other AQPs, AQP0 transports very little water when expressed in Xenopus oocytes which is mercury insensitive and its activation energy is ~7 kcal/mole. As a mutant AQP0, truncated at the carboxy-terminal, resulted in no increase in water permeability, AQP0 itself may transport water. The reconstitution of AQP0 in planar lipid bilayers resulted in the formation of a slightly anion-selective and voltage-dependent large channel. However, water permeabilty of these bilayers has not been critically examined. Interestingly, its ion channel activity was not observed when expressed in oocytes. Its physiological function should be related to keep lens transparency because a mutant AQP0 causes cataract in mice [26].

AQP1 is the first characterized water channel. It stimulates water transport when expressed in Xenopus oocytes, yeast vesicles, and mammalian cells, or when reconstituted into proteoliposomes. AQP1 was originally isolated from human red blood cells and expressed in wide variety of tissues; kidney, eye, spleen, intestine, lung, liver, heart, prostate, ovary, and choroid plexus epithelium. AQP1 is also expressed in vascular endothelia. In the eye, AQP1 is localized to corneal endothelium and ciliary body. In salivary gland, AQP1 is expressed at the basolateral membrane of acinus cells. In the kidney, it is expressed at apical and basolateral membranes of proximal tubules and thin descending limb of Henle loop. It is also expressed at endothelia of vasa recta. AQP1 is regarded as a constitutive water channel and its expression in the kidney was not regulated by dehydration or ADH. However, a

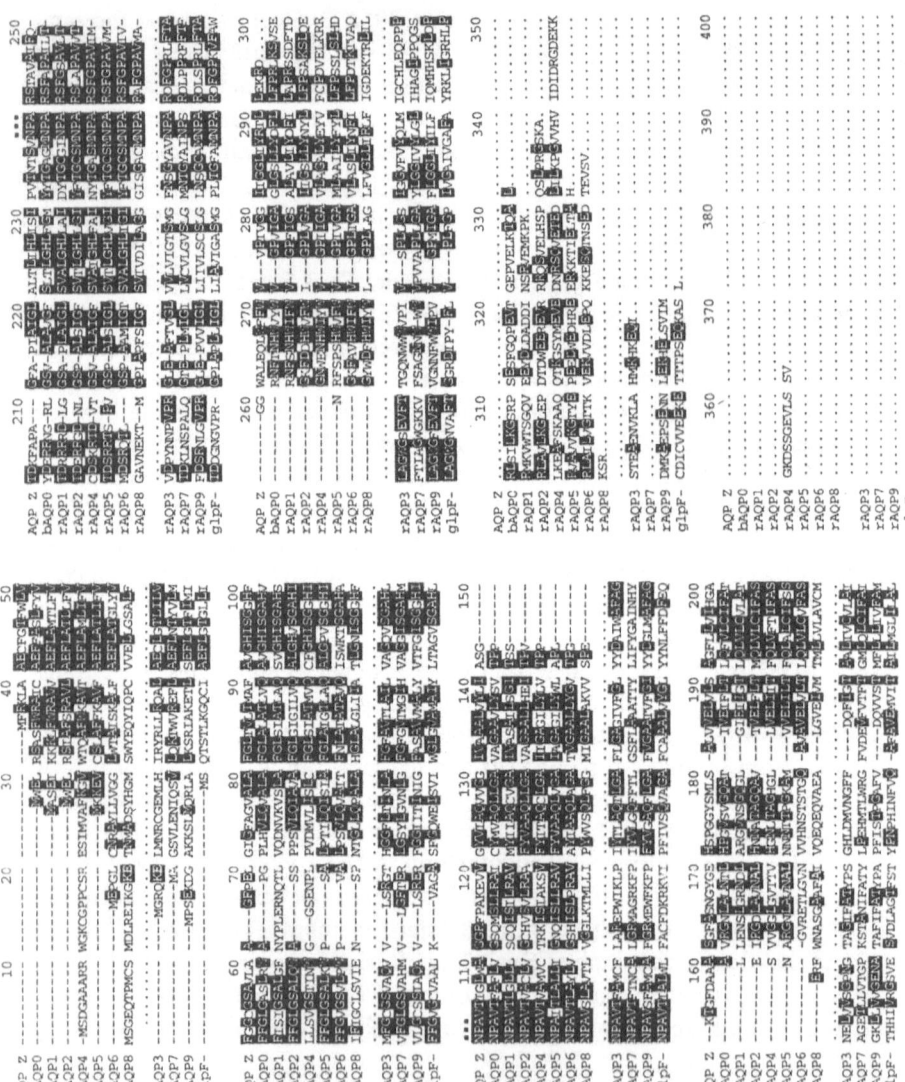

Fig. 4. Multiple sequence alignments of mammalian AQP members (AQP0–AQP9) and two AQP/MIP proteins of E. coli. (AQP Z and glpF). The upper 8 members belong to CHIP group and the lower 4 members belong to GLP group. "b" represents bovine and "r" represents rat. The NPA boxes are highlighted by three dots. Note that the two inserts are present in GLP group (144–159, 250–260). AQP8 is poorly conserved in CHIP group

Table 2. Characteristics of mammalian aquaporins

Characteristic	Aquaporin									
	0	1	2	3	4	5	6	7	8	9
Protein size (amino acid)	263	269	271	292	301/323	265	282	269	263	295
mRNA size (kb)	1.4	2.8	1.9	1.8	5.5	1.6	2.2	1.5	1.5	3.5
N-linked glycosylation	–	+	+	+	–	+	–	–	+	+
Hg inhibition of water transport	–	+	+	+	–	+	+	–	+	+
Glycerol transport	±	–	–	+	–	–	–	+	–	±
PKA site	+	–	+	–	+	+	+	–	–	+
PKC site	+	+	+	+	+	+	+	+	+	+
Induction by dehydration	?	–	++	+	–	?	+	–	–	?
Exon numbers	4	4	4	6	4	4	4	6	5	6
Gene locus	12q13	7q14	12q13	9q13	18q22	12q13	12q13	9p13	16p12	15q22

Hg, mercury; PKA, protein kinase A phosphorylation; PKC, protein kinase C phosphorylation; –, not present; ±,controversial; + present; ++,strong induction; ?, unknown.

K. Ishibashi et al.

Table 3. Tissue distribution of mammalian aquaporins

Characteristic	Aquaporin									
	0	1	2	3	4	5	6	7	8	9
Brain		+	-	+	++	-	-	-	-	+
Eye	+	+	-	+	+	+		±	-	+
Salivary gland		+	-	+	+	+		+	-	
Trachea		+		+	+	+		+	+	
Lung		+		±	+	++				+
Heart		+		-	+		-		-	-
Liver		+		+	-	-	-	+	+	+
Pancreas		-		+	+	-	-		+	-
Spleen		+		+	±	-	-		-	+
Small intestine		+		+	-	-		+	+	-
Colon		+		++	+			-	+	-
Kidney		+	++	++	+	+	+	+		+
Testis		-	±	-	-		·	++	++	+
Ovary		+			+	+		±		
Placenta		+		-	-				+	
Muscle		-		-	+					-
Leucocyte		-		±	-					++
Mammary gland					-	+		+		

-, absent; ±, very little; +, present; ++, abundant; blank, not known.

recent report of AQP1 in cholangiocyte showed that it is regulated by secretin [27]. The secretin-induced osmotic water permeability of isolated rat cholangiocyte was shown to be inhibited by mercury chloride. Secretin increased the expression of AQP1 at the plasma membrane with its proportional decrease at the intracellular membranes. The redistribution of AQP1 and the increase of membrane water permeability induced by secretin were both inhibited by the microtubule blocker, colchicine and by the exposure to a low temperature. These observations suggest that secretin induces the microtubule-dependent targeting of AQP1-containing vesicles to the plasma membrane. The water permeability by AQP1 has been shown to be stimulated by cAMP-induced phosphorylation although AQP1 itself does not have consensus phosphorylation site by protein kinase A [28]. The same authors also reported that AQP1 acquires ion channel activity when stimulated by protein kinase A. However, the results were not reproduced by other researchers. Much recent report shows that ADH stimulates and ANP inhibits AQP1 induced water permeability in oocytes [29]. These observations seem to favor a cAMP-dependent membrane shuttle mechanism for AQP1 regulation. However, these results should be confirmed in other systems. AQP1 was shown to be transiently induced by a growth factor, platelet-derived growth factor, in 3T3 fibroblastic cells. The need of water channel during cell division is not apparently evident. The cell swelling with accumulated water through water channel may facilitate cell division. People with defective AQP1 live normal life although their red blood cells have low water permeability. Knockout mice of AQP1 suffer from urine concentration defect. Further analysis of the knockout mice suggest the role of AQP1 in peritoneal fluid transport and intestinal fat absorption [5].

AQP2 is the first ADH-regulated water channel. AQP2 is localized at apical membrane of principle cells of collecting ducts. Water deprivation results in a drastic increase in mRNA and protein of AQP2. ADH modulates AQP2 in three ways [11]. First, ADH stimulates AQP2 gene transcription through cAMP responsive element (CRE) which is identified in the 5'flanking region of the AQP2 gene. Second, ADH induces redistribution of AQP2 from intracellular membranes to apical plasma membrane in a microtubule-dependent manner by protein kinase A phosphorylation at the carboxy-terminus, specifically at S256 whose mutation abolishes this trafficking. Third, ADH stimulates water permeability in protein kinase dependent manner other than the second mechanism when expressed in oocytes. Third observation, however, is controversial, because in other systems, isolated kidney collecting duct vesicles, the direct phosphorylation of AQP2 by protein kinase A did not affect its water permeability. Initially the expression of AQP2 was shown only in kidney. However, recently AQP2 is shown to be present in testis and inner ear [30]. As vasopressin is produced in testis, AQP2 may also be regulated in testis. As the role of vasopressin in Meniere's disease has been suggested, AQP2 may also have a role in endolymph formation. However, people with defective AQP2 only suffer from polyuria, a disease named nephrogenic diabetes insipidus. The roles of AQP2 in inner ear and testis remain to be clarified.

AQP3 is the first functionally unique water channel; it transports glycerol as well as water. Although AQP0, 1, and 2 has been shown to transport glycerol by some investigators, others constantly did not observe such transports. Three independent discoverers of AQP3 agreed on its glycerol permeability. The issue of water permeability of AQP3 seems to be resolved when Verkman's group reported the comparison of five AQPs water permeabilities in oocyte system [31]. AQP3 is expressed at the basolateral membrane of principle cells of collecting ducts. The expression is mainly distributed from the outer medulla to the cortex in rat kidney. AQP3 expression at inner medulla gradually decreases where AQP4 is mainly expressed. In colon, AQP3 is expressed at the basolateral membrane of columnar epithelial cells at the surface. AQP3 is mainly expressed in distal colon. In Northern blot, AQP3 is expressed widely in GI tracts [32]. However, the immunohistochemistry revealed its presence only at the distal colon. This discrepancy of expression between RNA and protein levels may suggest the presence of homolog(s) of AQP3 in GI tracts which reacts with AQP3 cDNA probe but not with AQP3 antibody which is made by immunizing C-terminus peptides of AQP3. Our initial report of AQP3 showed the presence of AQP3 mRNA at spleen. Recently, AQP3 has been shown to be present at red blood cells [33]. Red blood cells seem to have at least two AQPs, AQP1 and 3. Although AQP3 seems to be constitutively expressed at basolateral membrane of collecting ducts, its expression is induced by dehydration in kidney and by glucocorticoid in pulmonary cell lines [34].

AQP4 is the first mercurial-insensitive water channel. This property is not unique in AQP4 as AQP7 is also mercurial-insensitive. AQP4 is predominantly expressed in brain, and inside the brain it is strongly expressed in cells lining the ventricular surface. These cells are not neurons but glial cells and some of them may be neural stem cells. The primary structure of AQP4 is similar to big brain (Bib) of Drosophila. The loss of function of Bib produces twice as big brain in fly because Bib may be important for the suppression of neural differentiation. Without Bib function, all neural epithelial cells differentiate to neurons, although the underlying mechanism is unclear. As the knockout mouse of AQP4 did not show any gross neurological defect in morphology and physiology [5], AQP4 may not be the Bib ortholog or the role of AQP4 has changed in mammals. AQP4 in mammals may contribute to reabsorption and maintenance of cerebrospinal fluid. AQP4 is also expressed at the basolateral membrane of tracheal and bronchial epithelia, but not in alveolar epithelium. The AQP4 expression in kidney is restricted to the basolateral membrane of inner medullary collecting duct cells. AQP3 and 4 are colocalized in some epithelial membranes such as collecting duct cells, colonic epithelia, and tracheal epithelia. Such redundant expression of AQPs in the same membrane domain is intriguing, but found in several tissues. For example, AQP1 and AQP7 are colocalized at brush border membrane of proximal tubules. AQP2 and AQP7 are colocalized at the tail of late spermatids. AQP1 and AQP3 are colocalized at red blood cells. Given that the function of AQPs is critical, such coexpression may suggest the safeguard against the loss of one of aquaporins to keep water permeability high. The reason for the relatively minor phenotypic changes of knockout animals and defective humans of aquaporins may lie in such redundant expressions of AQPs. AQP4 is also unique in that its gene yields two distinct mRNA species corresponding to two polypeptides, which differ at their amino-termini. Both are functional aquaporins and expressed in the same tissues suggesting the presence of the heterotetramers. On the other hand, the spliced-out shorter form of AQP4 without water channel function was reported in the original report, which was not confirmed by other researchers. Mice with defect in AQP4 only have minor defect in urine concentrating ability. The roles of AQP4 in muscle, brain, and colon remain to be clarified.

AQP5 is expressed at apical membranes in exocrine gland. As AQP5 is more closely related to AQP2 and have a protein kinase A consensus phosphorylation site, it is expected that AQP5 molecule be translocated from intracellular vesicles to the apical membrane by neurohormonal stimulation. However, it was not the case with AQP5 in the lacrimal gland [35]. Although the tear secretion is stimulated by pilocarpine, the expression level of AQP5 on the apical membrane in the pilocarpine-stimulated lacrimal glands was not increased. However, carboxy terminus region seemed to be modified (masked or altered structure) by pilocarpine which was suggested by the change in the immunohistological staining by the caboxy terminus-recognizing antibody. The mechanism underlying the increase of water transport through AQP5 by neurohumoral stimulation may be a novel regulatory mechanism of aquaporins, which may be used by other AQPs. Recent AQP5 knockout mice revealed decreased fluid secretion of salivary glands [36].

AQP6 stimulate water transport very weakly when expressed in Xenopus oocytes. AQP6-stimulated water transport was inhibited by mercury chloride. AQP6 is selectively expressed in the kidney, similarly in the cortex and the medulla. It is localized at podocytes in glomeruli and at the endosomes of proximal tubules and type A intercalated cells in collecting ducts [37]. The dehydration and alkalosis induced its expression. As AQP6 is localized at acid secreting cells and colocalized with H-pump, the possible role of AQP6 in acid-base regulation should be seriously considered. It is possible that small water channel activity of AQP6 is due to the absence of surface membrane targeting in oocyte expression system. However, a similarly intracellularly localized AQP-TIP (tonoplast aquaporins of plant) can function as water channel in oocytes. The low water channel activity of AQP0 and AQP6 may possibly be enhanced by some unknown stimulatory factors or procedures.

AQP7 was cloned from testis and stimulates urea and glycerol permeability as well as water that are mercury-insensitive. Although rat AQP7 has a very short carboxy-terminus, human and mouse AQP7 have long ones [38]. Such big species difference of AQPs is unusual. The examination of its genomic structure revealed that this difference couldn't be explained by alternative splicing. The possibility of two highly related genes cannot be ruled out. AQP7 is predominantly expressed in testis, where AQP7 is localized to mid and tail portion of late spermatids [39]. AQP7 is also expressed in mature sperm and may correspond to the functional water channel observed in human sperms, which is mercury-insensitive. As the osmolarity of seminiferous tubules is hypertonic (385 mOsm/kg), water channels of late spermatids may facilitate the shrinking process of cytoplasm, especially tail portion which is otherwise bulky. Recently, the role of glycerol facilitator of yeast in cell fusion during mating has been reported [16]. If the osmotic state of cell can regulate cell fusion, the role of AQP7 may lie in the process of sperm-egg fusion. AQP7 is also expressed at the brush border membrane of proximal tubules in kidney. The presence of AQP7 in adipose tissue has been documented [40]. As AQP7 transport glycerol, its role in lipid metabolism is worthy of further investigation.

AQP8 is a water selective aquaporin and its primary structure is unique among other aquaporins especially in amino terminus half. Its carboxy-terminus half is similar to plant AQP-TIP which is localized to intracellular tonoplast membranes. Therefore, the possibility that AQP8 is present at intracellular membrane as shown with AQP6 should be examined. It is most abundantly expressed in testis. It seems that testis has at least four aquaporins, AQP2, 7, 8, and 9. This unexpected complexity of aquaporins in testis should calls for the further analysis of the role of aquaporins in reproduction biology. AQP8 is also expressed in liver, pacrease, placenta, small intestine, and salivary gland in small amount. In situ hybridization showed its expression at hepatocytes and gland of pancreas. The functional study reported the absence of functional water channel in isolated hepatocytes. It is possible that AQP8 is localized at intracellular membranes or at the limited membrane of biliary intercellular canaliculi of hepatocytes. Alternatively, AQP8 may also transport small solutes as mouse AQP8 transports urea as well as water but not glycerol [21]. However, rat and human AQP8 transport only water [25, 41, 42]. This functional discrepancy remains to be resolved.

AQP9 is the latest water channel that also permeates urea. It is cloned from liver and may function as an urea transporter. As liver is the main producer of urea, the presence of some urea exit mechanism from hepatocyte is suspected. AQP9 is the good candidate for this urea exit pathway. The human AQP9 did not permeate glycerol similar to AQP-CE1 (a C. elegans aquaporin)[22]. However, rat AQP9 permeates wide variety of non-charged solutes including carbamides, polyols, purines, and pyrimidines [43, 44]. The pore size of rat AQP9 seems to be larger than that of AQP3 suggesting that AQP9 may play an important role in hepatocyte to transport metabolites. The human but not rat AQP9 is abundantly expressed in peripheral leukocytes. Lung and spleen but not thymus express AQP9 in small amount. The functional water channel of leukocytes was not identified in previous studies. The possible role of AQP9 in immunological function of leukocytes needs to be investigated.

3
Molecular Structure of AQP/MIP Family

3.1
Oligomerization

Biochemical analysis to estimate the molecular weight of MIP family indicated that the native proteins exist as homo-tetramers in the membrane. The detailed analysis of AQP1 extracted from red cell membranes showed that AQP1 is in a form of homo-tetramer with one glycosylated and three unglycosylated subunits. Unlike other membrane proteins, all AQP1 proteins exist in tetramer, and not a mixture of monomers and dimers. Electron microscopy analysis of AQP1 and AQP0 supported the tetrameric structure of these proteins.

The above results imply that all AQP/MIP family members are in tetramer. However, a previous report on the glycerol facilitator of E. coli (GlpF) suggested that it might exist as a homodimer. Furthermore, a recent report by Lagree et al. revealed that GlpF exists as a monomer [45]. The experimental method they used is the sucrose gradient sedimentation after solubilization in the nondenaturing detergent (n-octyl b-D-glucopyranoside,

OG) or denaturing detergent (SDS). An insect AQP, AQPcic that belongs to CHIP group sedimented to the position corresponding to the molecular weight of tetramer when solubilized in OG, but to that equal to monomer when solubilized in SDS. On the contrary, GlpF sedimented at the monomer position even when solubilized in OG. They further showed that introducing mutations at the 6th transmembrane domain of AQPcic to simulate to that of GlpF made AQPcic permeable to glycerol and lost its water permeability [46]. This mutated AQPcic no longer existed as tetramer but as monomer in OG. These results suggest that a monomer form will function as glycerol channel. The function of aquaglyceroporin may be explained by the mixture of tetramer form (water permeable) and monomer form (glycerin permeable). However, a monomer-forming mutant of AQP2 has been shown to be trapped in endoplasmic reticulum [47]. The analysis of members of GLP group that transport water and glycerol like AQP3 is necessary.

The importance of the oligomerization on channel regulation was implicated in mutation-phenotype analysis of human nephrogenic diabetus insipidus patients [47]. This disease is caused by mutations of AQP2 that belongs to CHIP group, and two inheritance forms; recessive and dominant forms are known. Similar sedimentation studies showed that the AQP2 protein created by the recessive mutation migrated at a monomer position. Wild type AQP2 migrated at the position of tetramer. Interestingly, the mutation causing a dominant inheritance seemed to form tetramers with wild type AQP2. These results are more informative if interpreted together with trafficking events of AQP2 inside the cells. The wild-type AQP2 were correctly expressed at the plasma membrane of the oocytes, whereas the recessive AQP2 mutants were retarded in endoplasmic reticulum (ER) and the dominant mutant was retained in Golgi compartment. Thus, the recessive mutants would be retained in ER due to misfolding and/or failure to form tetramers. Normal AQP2 produced from normal allele would form tetramers and reach the plasma membrane, which explains the recessive inheritance. Because the dominant mutant can form tetramers with normal and mutated proteins, three quarters of the tetramers may contain mutated subunits. The presence of mutant subunits in the tetramers may be a signal for the entrapment in Golgi compartment, and this dominant-negative effect explains a dominant inheritance pattern. The mechanism for mutant-containing tetramers to fail to reach plasma membrane from Golgi complex is currently unknown.

In potassium channels, the ion-permeating pore is formed by homotetramer of subunits. In contrast, each AQP1 monomer functions as individual water channel although it exists as tetramer. Coexpressing AQP1 with mutated AQP1 (nonpermeable mutants or mercury insensitive mutants) in

oocytes showed that each monomer functions independently and no dominant negative effect by inactivated mutants were observed. The dominant
negative effect of AQP2 mutant is caused by the impaired routing and not by
the loss of channel activity [47]. These results together with the monomeric
function of glycerin transporter suggest that AQP/MIP proteins function
independently, i.e. each monomer can work as a water or glycerol channel.
However, the pore structure may be modulated by the state of subunits;
monomer or tetramer.

3.2
Hourglass Model

Agre and coworkers proposed the hourglass model for the structure of
AQP1 [48]. The basic idea is that the highly conserved two NPA motifs will
bend into the membrane to form a central complex in the membrane. The
NPA motifs are relatively hydrophobic and buried in the membrane. This
complex will form an aqueous pathway, a pore. This model was based on the
studies showing the importance of loop B and E in water channel function.
They observed that A73 and C189 located close to the NPA motifs are important for the inhibition by mercurial agents. This model explains the necessity
of the conservation of NPA boxes. However, the exact structure how these
two motifs interact with each other and with membrane spanning helices
remains to be clarified.

We also performed structure-function studies of AQP2 in oocytes expression system [49]. A similar topology as AQP1 was confirmed by introducing the N-glycosylation signal (NTS motif) into the loops. In our analysis, the importance of C181 corresponding to C189 of AQP1 was confirmed,
however, A65 corresponding to A73 of AQP1 was not important for mercurial inhibition. Instead, the importance of the loop C and D, in addition to
loop E, was observed. The result suggests that loop C and D may also contribute to the formation of the aqueous pore in AQP2 protein. This discrepancy will be also solved when more minute structures of AQP/MIP proteins
are resolved by direct visualization methods (see below).

3.3
Direct Visualization

The above model can be more directly examined by direct visualization of
membrane proteins. The electron microscopy applied to native aquaporin-
rich membranes revealed the presence of tetramer-like structures. Technol-
ogy and methodology have advanced in this field as well, and it was essential

to prepare a pure protein in mg order and to reconstitute it into lipid bilayer, so that the projection of two-dimensional crystal is obtained. So far, this approach has been successfully applied to AQP1 and AQP0 [50] as these proteins were easily obtained from native tissues. Walz et al. reconstituted AQP1 from red blood cells into lipid bilayer at a high protein-lipid ratio, and examined the resulting 3 μm diameter membrane vesicles by high-resolution cryoelectron microscopy. The projection of stained two-dimensional crystals clearly showed the tetramer structure of AQP1. His group and other two groups independently further refined the analysis and identified the presence of 6–8 -helices in each monomer. In 1997, the resolution was ~6 Å in xy direction and 6–20 Å in z direction. They all agreed that each monomer consists of six highly tilted transmembrane -helices surrounding a central density. This central density was assigned to be loop B and E that bend into the membrane. This picture is a morphological demonstration of the hourglass model. This central complex together with surrounding helices could form an aqueous pore, which selectively permeates water at a high rate. Clearly, the atomic resolution is needed for the final proof of this model, and this will be done in the near future by X-ray crystallography applied to three-dimensional crystals as has been successfully done for a potassium channel of Streptomyces lividans.

Other than AQP1 and AQP0, very few studies have been conducted for the structural analysis. The members of GLP group in particular should be examined to clarify the structural basis for their functional difference.

4
Function of AQP/MIP Family

Not all functions of AQP/MIP family members have been clarified and new functions of this family will be identified in the near future. We will review the currently known several functions of this family. Some are controversial due to the different expression systems and/or the different methods for detection.

4.1
Water Channel

When AQP1 was expressed in oocytes, the known biophysical characteristics of water channel were all observed. (1) High osmotic water permeability coefficient (Pf). Pf is the net water flow of volume across the membrane in response to a hydrostatic or osmotic driving force. The Pf larger than 0.01 cm/s suggests the presence of water channels. The diffusional water permeability (Pd) is the water movement without osmotic gradient and can be measured by tracer fluxes. Pf/Pd larger than unity indicates the presence of water channels. The value of Pf/Pd was interpreted to be the number of water molecules present in the narrow aqueous pathway where water moves as a single file. (2) Low activation energy (Ea). The bulk flow of water through water channels makes the water transport little dependent on temperature. Water transfer across the membrane by simple diffusion shows Ea>10 kcal/mol, and Ea less than that (typically less than 6 kcal/mol) strongly suggests the water flow through a pore. (3) Inhibition by sulphdryl reagents such as mercurials. These are not specific inhibitors for water channels and they should be used in lower concentration (less than 1 mM and preferably at 0.3 mM in oocytes) to avoid nonspecific inhibition. The reversal of its effect by β-mercaptoethanol is often used to show the non-toxic effect of mercurial. Some of AQPs such as AQP4 and AQP7 have been shown to be mercury insensitive. The mercury sensitive cysteines or methionines may give clues to the pore segments. In the case of AQP1 and AQP2, the mercury sensitive cysteine is located in loop E. Similar but not identical result was obtained with AQP4. In the case of AQP3, the incorporation of a cysteine in loop E made water and glycerol transport more sensitive to mercury, suggesting the pore structures of CHIP group and GLP group may be similar [51]. Interestingly, plant AQP-TIP has different mercury sensitive cysteine. It is located at third transmembrane segment [52]. (4) Exclusion of other solutes and ions. The bulk water flow is observed with large-pore forming proteins such as amphotericin B and the bacterial channel hemolysin [53]. However, these proteins also permeate ions. Water channels exclude ions suggesting that the pore is a single file structure.

Whether all the members of AQP/MIP family function as water channel is a challenging problem. Only GlpF has been shown to have no water channel activity either in oocyte expression system or in the native state. Other examined members seem to have water channel activities.

4.2
Small Solute Channel

Most of CHIP proteins do not permeate small solutes such as glycerol and urea. AQP1 was shown to be water selective in oocyte system and subsequent studies with the liposome system confirmed this conclusion. However, nodulin 26 has been shown to transport water and glycerol when incorporated in liposomes [13]. Furthermore, some of the members of plant MIP family (PIP and TIP) also permeate glycerol as well as water [18, 19]. Therefore, some members of CHIP group may permeate small solutes. The glycerol permeation by AQP1 was shown by some investigators. This controversy has not yet been resolved. On the other hand, the members of GLP group usually permeate glycerol. In the case of AQP3, both water and glycerol permeabilities are phloretin or mercury sensitive and show low activation energy. Although the inhibition is relatively weak compared with that in AQP1, the introduction of a cysteine residue in loop E made AQP3 more sensitive to mercury in both water and glycerol transports [51]. Therefore, water and glycerol may permeate the same pore. However, other investigators reported the absence of inhibition by mercury in the case of glycerol permeation that had higher activation energy, suggesting the separate pathway for water and glycerol [54]. In other members of GLP group, channel like mechanism of glycerol transport has been demonstrated. The transport of glycerol through GlpF has low temperature dependence (4.5 kcal/mol) and is inhibited by mercury and phloretin. The solutes and water permeability were also inhibited by phloretin and mercury in rat AQP9 [43]. The glycerol transport of one of the yeast GLP member (FPS1) has been shown to have low activation energy [55]. Thus, these data indicate that solute transports through GLP proteins are pore-mediated.

FPS1 seems to function as a glycerol exit pathway to balance the osmotic difference between cytosol and extracellular environment [56]. Therefore, the defect of FPS1 was overcome by 1M sorbitol in the medium. FPS1 is important for cell fusion of yeast and cell assimilation to hypotonic medium after exposing to hypertonic environment [16]. FPS1 seems to be closed or endocytosed at hyperosmotic condition to keep glycerol inside the cell. As yeast uses glycerol as an osmolyte (compatible solute to protect cells from hypertonic environment), FPS1 as a glycerol channel may play an important role. In some species, sorbitol is used as one of osmolytes. As rat AQP9 permeates manitol and sorbitol [43], some of GLP members may function as sorbitol channels. The identification of such physiologically relevant solutes is important to clarify the role of GLP members.

4.3
CO2 Gas Channel

Plasma membranes have a high gas permeability that is mediated by dissolving of gas into the membrane lipid. Interestingly, some membranes such as oocytes have low gas permeability. Therefore, the increase of CO_2 permeability was detectable when AQP0-5 and nodulin 26 were expressed in oocytes [57]. In the case of AQP1 expression, CO_2 permeability was inhibited by mercury, Zn, and DIDS, whereas water permeability was only inhibited by mercury. Therefore, the permeation mechanism of water and CO_2 may be different. Because plasma membranes generally have high gas permeability, the physiological significance of CO_2 permeability and possible permeability of other gases (oxygen and NO) by AQP/MIP family, remains to be clarified.

4.4
Metal Channels

Glycerol facilitator of E.coli (GlpF) has been shown to transport antimonite [58]. The antimonite seems to be transported in non-ionic form $(Sb(OH)3)$ which is recognized as a polyol by GlpF. The inactivation of GlpF resulted in acquiring the resistance against antimonite, suggesting that GlpF is the major route of entry of animonite into cells. Another example of possible metal transport by AQP/MIP family is the member of CHIP group (smpX) in Synechococcus. The deletion of smpX made the bacteria hypersensitive to copper, suggesting that smpX may be involved in copper homeostasis to which an ABC transporter is implicated [59]. The exact role of smpX in copper transport remains to be clarified.

4.5
Regulation of Transport

The gating of water channels has not been seriously considered because water fluxes always follow the osmotic gradient and no directionality of water flow is observed. The osmotic gradient is formed and regulated by ion and solute channels and transporters. So the focus of research is more shifted to the regulation of such channels and transporters. However, recent identification of solutes and water transport by AQP/MIP family suggests the possible regulation of transport by these proteins. As the method of excised patch clamping type is not applicable to AQP/MIP proteins in the plasma membrane, channel gating and exocytotic insertion of channels are difficult to distinguish. The phosphorylation has been shown to stimulate the water transport in AQP2, AQP5, AQP-PIP, and AQP-TIP, and the solute transport in nodulin 26. The serine residues to be phoshorylated are localized in N-and C-termini in the cytoplasm. If such stimulation is caused by gating of the channel pore, ball and socket type of regulation may be involved as shown in potassium channels and chloride channels. Recently, acidic external pH has been shown to inhibit water and glycerol permeation through AQP3 [60]. The result was specific to AQP3 and AQP0, 1, 2, 4, and 5 were not inhibited by acidic pH. Interestingly, pK for water channel inhibition was higher than pK for glycerol channel. Therefore, at low pH, AQP3 may function predominantly as glycerol channel. The pH sensitivity of water and glycerol permeation suggests that water and glycerol permeate through AQP3 by forming successive hydrogen bonds with titrable sites. The presence of free -OH will facilitate to form hydrogen bonds. Although glycerol is larger than urea, glycerol is more efficiently transported than urea in AQP3. This result may be explained by this mechanism. Whether this mechanism is correct and is also present in the other members of GLP group remains to be clarified.

5
Perspectives

The functional studies of AQP/MIP family are well behind the identification of new members of this family. Many uncharacterized members of AQP/MIP family are present. The care should be taken not to be biased by the primary sequence because some members of GLP group may not transport glycerol (AQP-CE1, hAQP9), and some members of CHIP group may transport urea (mAQP8, nodulin 26). At the same time the members of AQP/MIP family with low homology to the current members should be searched for. The clue for water selective pore will be identified by comparing the primary sequences among extended AQP/MIP family, which in turn help to identify novel water channels irrelevant to AQP/MIP family. In the mean time, the physiological significance of each member of AQP/MIP family should be clarified by developing specific inhibitor or stimulator, or by analyzing the knockout mice. There is also relatively little information on the regulation of AQP/MIP family except AQP2. The identification of the mechanisms of genetic and posttranslational regulation of AQP/MIP family is also necessary. The clarifying the hormonal and metabolic regulation of AQP/MIP family and the interaction with other channels and transporters should broaden our understanding of water and solutes transport in the living organisms.

References

1. Lee MD, King LS, Agre P (1997) The aquaporin family of water channel proteins in clinical medicine. Medicine 76:141–156
2. Yamamoto T, Sasaki S (1998) Aquaporins in the kidney: Emerging new aspects. Kidney Int 54:1041–1051
3. Ma T, Verkman AS (1999) Aquaporin water channels in gastrointestinal physiology. J Physiol 517:317–326
4. Ishibashi K, Sasaki S (1997) Aquaporin water channels in mammals. Clin Exp Nephrol 1:247–253
5. Verkman AS (1999) Lessons on renal physiology from transgenic mice lacking aquaporin water channels. J Am Soc Nephrol 10:1126–1134

6. Agre P, Bonhivers M, Borgnia MJ (1998) The aquaporins, blueprints for cellular plumbing systems. J Biol Chem 273:14659–14662
7. Maurel C (1997) Aquaporins and water permeability of plant membranes. Annu Rev Plant Physiol Plant Mol Biol 48:399–429
8. Park JH, Saier MH (1996) Phylogenetic characterization of the MIP family of transmembrane channel proteins. J Membrane Biol 153: 171–180
9. Figeri A, Nicchia GP, Verbavatz JM, Velenti G, Svetlo M (1998) Expression of aquaporin-4 in fast-twitch fibers of mammalian skeletal muscle. J Clin Invest 102:695–703
10. Ishibashi K, Sasaki S (1998) The dichotomy of MIP family suggests two separate origins of water channels. News Physiol Sci 13:137–142
11. Sasaki S, Ishibashi K, Marumo F (1998) Aquaporin-2 and -3: Representatives of two subgroups of the aquaporin family colocalized in the kidney collecting duct. Annu Rev Physiol 60:199–220
12. Froger A, Tallur B, Thomas D, Delamarche C (1998) Prediction of functional residues in water channels and related proteins. Protein Sci 7:1458–1468
13. Dean RM, Rivers RL, Zeidel ML, Roberts DM (1999) Purification and functional reconstitution of soybean nodulin 26. An aquaporin with water and glycerol transport properties. Biochemistry 38:347–353
14. Delamarche C, Thomas D, Rolland JP, Froger A, Gouranton J, Svelto M, Agre P, Calamita G (1999) Visualization of AqpZ-mediated water permeability in Escherichia coli by cryoelectron microscopy. J Bacteriol 181:4193–4197
15. Bonhivers M, Carbrey JM, Gould SJ, Agre P (1998) Aquaporins in Saccharomyces. Genetic and functional distinctions between laboratory and wild-type strains. J Biol Chem 273:27565–27572
16. Jennifer P, Herskowitz I (1997) Osmotic balance regulates cell fusion during mating in Saccharomyces cervisiae. J Cell Biol 138:961–974
17. Weig A, Deswarte C, Chrispeels MJ (1997) The major intrinsic protein family of Arabidopsis has 23 members that form three distinct groups with functional aquaporins in each group. Plant Physiol 114:1347–1357
18. Gerbeau P, Guclu J, Riproche P, Maurel C (1999) Aquaporin Nt-TIPa can account for the high permeability of tobacco cell vacuolar membrane to small neutral solutes. Plant J 18:577–587
19. Biela A, Grote K, Otto B, Hoth S, Hedrich R, Kaldenhoff R (1999) The Nicotiana tabacum plasma membrane aquaporin NtAQP1 is mercuty-sensitive and permeable for glycerol. Plant J 18:565–570
20. Kuwahara M, Ishibashi K, Gu Y, Terada Y, Kohara Y, Marumo F, Sasaki S (1998) A water channel of the nematode C. elegans and its implications for channel selectivity of MIP proteins. Am J Physiol 275:C1459–C1464
21. Ma T, Yang B, Verkman AS (1997) Cloning of a novel water and urea-permeable aquaporin from mouse expressed strongly in colon, placenta, liver, and heart. Biochem Biophys Res Commun 240:324–328
22. Ishibashi K, Kuwahara M, Gu Y, Tanaka Y, Marumo F, Sasaki S (1998) Cloning and functional expression of a new aquaporin (AQP9) abundantly expressed in the peripheral leukocytes permeable to water and urea, but not to glycerol. Biochem Biophys Res Commun 244:268–274
23. Abrami L, Gobin R, Berthonaud V, Thanh HL, Chevalier J, Riproche P, Verbavatz JM (1997) Localization of the FA-CHIP water channel in frog urinary bladder. Eur J Cell Biol 73:215–221

24. Schreiber R, Greger R, Kunzelmann K (1999) CFTR activates endogenous AQP3 in Xenopus oocytes. Pflugers Arch 437:R91
25. Koyama N, Ishibashi K, Kuwahara M, Inase N, Ichioka M, Sasaki S, Marumo F (1998) Cloning and functional expression of human aquaporin8 cDNA and analysis of it gene. Genomics 54:169–172
26. Agre P (1998) Aquaporin null phenotypes: the importance of classical physiology. Proc Natl Acad Sci USA 95:9061–9063
27. Marinelli RA, Tietz PS, Pham LD, Rueckert L, Agre P, LaRusso NF(1999) Secretin induces the apical insertion of aquaporin-1 water channels in rat cholangiocytes. Am J Physiol 276:G280–286
28. Yool, AJ, Stamer WD, Regan JW (1996) Forskolin stimulation of water and cation permeability in aquapoin 1 water channels. Science 273:1216–1218
29. Patil RV, Han Z, Wax MB (1997) Regulation of water channel activity of aquaporin 1 by arginine vasopressin and atrial natriuretic peptide. Biochem Biophys Res Commun 238(2):392–396
30. Beitz E, Kumagami H, Krippeit-Drews P, Ruppersberg JP, Schultz JE (1999) Expression pattern of aquaporin water channels in the inner ear of the rat. The molecular basis for a water regulation system in the endolymphatic sac. Hear Res 132:76–84
31. Yang B, Verkman AS (1997) Water and glycerol permeabilities of aquaporins 1-5 and MIP determined quantitatively by expression of epitope-tagged constructs in Xenopus oocytes. J Biol Chem 272:16140–16146
32. Koyama Y, Yamamoto T, Tani T, Nihei K, Kondo D, Funaki H, Yaoita E, Kawasaki K, Sato N, Hatakeyama K, Kihara I (1999) Expression and localization of aquaporins in rat gastrointestinal tract. Am J Physiol 276:C621–627
33. Roudier N, Verbavatz JM, Maurel C, Ripoche P, Tacnet F (1998) Evidence for the presence of aquaporin-3 in human red blood cells. J Biol Chem 273:8407-8412
34. Tanaka M, Inase N, Fushimi K, Ishibashi K, Ichioka M, Sasaki S, Marumo F (1997) Induction of aquaporin 3 by corticosteroid in a human airway epithelial cell line. Am J Physiol 273:L1090–1095
35. Ishida N, Hirai SI, Mita S (1997) Immunolocalization of aquaporin homologs in mouse lacrimal glands. Biochem Biophys Res Commun 238:891–895
36. Ma T, Song Y, Gillespie A, Carlson EJ, Epstein CJ, Verkman AS (1999) Defective secretion of saliva in transgenic mice lacking aquaporin-5 water channels. J Biol Chem 274:20071–20074
37. Yasui M, Kwon TH, Knepper MA, Nielsen S, Agre P (1999) Aquaporin-6: An intracellular vesicle water channel protein in renal epithelia. Proc Natl Acad Sci USA 96:5808–5813
38. Ishibashi K, Yamauchi K, Kageyama Y, Saito-Ohara F, Ikeuchi T, Marumo F, Sasaki S (1998) Molecular characterization of human Aquaporin-7 gene and its chromosomal mapping. Biochim Biophys Acta 1399:62–66
39. Suzuki-Toyota F, Ishibashi K, Yuasa S (1999) Immunohistochemical localization of a water channel, aquaporin 7 (AQP7), in the rat testis. Cell Tissue Res 295:279–285
40. Kuriyama H, Kawamoto S, Ishida N, Ohno I, Mita S, Matsuzawa Y, Matsubara K, Okubo K (1997) Molecular cloning and expression of a novel human aquaporin from adipose tissue with glycerol permeability. Biochem Biophys Res Commun 241(1):53–58

41. Ishibashi K, Kuwahara M, Kageyama Y, Tohsaka A, Marumo F, Sasaki S(1997) Cloning and functional expression of a second new aquaporin abundantly expressed in testis. Biochem Biophys Res Commun 237:714–718

42. Koyama Y, Yamamoto T, Kondo D, Funaki H, Yaoita E, Kawasaki K, Sato N, Hatakeyama K, Kihara I (1997) Molecular cloning of a new aquaporin from rat pancreas and liver. J Biol Chem 272:30329–30333

43. Tsukaguchi H, Shayakul C, Berger UV, Mackenzie B, Devidas S, Guggino WB, van Hoek AN, Hediger MA (1998) Molecular characterization of a broad selectivity neutral solute channel. J Biol Chem 273:24737–24743

44. Ko SB, Uchida S, Naruse S, Kuwahara M, Ishibashi K, Marumo F, Hayakawa T, Sasaki S (1999) Cloning and functional expression of rAOP9L a new member of aquaporin family from rat liver. Biochem Mol Biol Int 47:309–18

45. Lagree V, Froger A, Deschamps S, Pellerin I, Delamarche C, Bonnec G, Gouranton J, Thomas D, Hubert JF (1998) Oligomerization state of water channels and glycerol facilitators. Involvement of loop E. J Biol Chem 273:33949–33953

46. Lagree V, Froger A, Deschamps S, Hubert JF, Delamarche C, Bonnec G, Thomas D, Gouranton J, Pellerin I (1999) Switch from an aquaporin to a glycerol channel by two amino acids substitution. J Biol Chem 274:6817–6819

47. Kamsteeg EJ, Wormhoudt TA, Rijss JP, van Os CH, Deen PM (1999) An impaired routing of wild-type aquaporin-2 after tetramerization with an aquaporin-2 mutant explains dominant nephrogenic diabetes insipidus. EMBO J 18:2394–3400

48. Mathai JC, Agre P(1999) Hourglass pore-forming domains restrict aquaporin-1 tetramer assembly. Biochemistry 38:923–928

49. Bai L, Fushimi K, Sasaki S, Marumo F (1996) Structure of aquaporin-2 vasopressin water channel. J Biol Chem 271:5171–5176

50. Hasler L, Walz T, Tittmann P, Gross H, Kistler J, Engel A(1998) Purified lens major intrinsic protein (MIP) forms highly ordered tetragonal two-dimensional arrays by reconstitution. J Mol Biol 279:855–864

51. Kuwahara M, Gu Y, Ishibashi K, Marumo F, Sasaki S (1997) Mercury-sensitive residues and pore site in AQP3 water channel. Biochemistry 36:13973–13978

52. Daniels MJ, Chaumont F, Mirkov TE, Chrispeels MJ (1996) Characterization of a new vacuolar membrane aquaporin sensitive to mercury at a unique site. Plant Cell 8:587–599

53. Paula S, Akeson M, Deamer D (1999) Water transport by bacterial channel a-hemolysin. Biochim Biophys Acta 1418:117–126

54. Echevarria M, Windhager EE, Frindt G (1996) Selectivity of the renal collecting duct water channel aquaporin-3. J Biol Chem 271:25079–25082

55. Coury LA, Hiller M, Mathai JC, Jones EW, Zeidel ML, Brodsky JL(1999) Water Transport across Yeast Vacuolar and Plasma Membrane-Targeted Secretory Vesicles Occurs by Passive Diffusion. J Bacteriol 181:4437–4440

56. Tamas MJ, Luyten K, Sutherland FC, Hernandez A, Albertyn J, Valadi H, Li H, Prior BA, Kilian SG, Ramos J, Gustafsson L, Thevelein JM, Hohmann S (1999) Fps1p controls the accumulation and release of the compatible solute glycerol in yeast osmoregulation. Mol Microbiol 31:1087–1104

57. Boron WF, Cooper GJ (1999) Gas channels. Pflugers Arch 437:R41

58. Sanders OI, Rensing C, Kuroda M, Mitra B, Rosen BP (1997) Antimonite is accumulated by the glycerol facilitator GlpF in Escherichia coli. J Bacteriol 179:3365–3367

59. Kashiwagi S, Kanamuru K, Mizuno T(1995) A Synechococcus gene encoding a putative pore-forming intrinsic membrane protein. Biochim Biophys Acta 1237:189–192
60. Zeuthen T, Klaerke DA (1999) Transport of water and glycerol in aquaporin 3 is gated by H. J Biol Chem 274:21631–21636

The Mechanisms of Aquaporin Control in the Renal Collecting Duct

E. Klussmann[1], K. Maric[1], and W. Rosenthal[1,2]

[1] Forschungsinstitut für Molekulare Pharmakologie, Alfred-Kowalke-Straße 4, D-10315 Berlin, Germany.
[2] Freie Universität Berlin, Institut für Pharmakologie, Thielallee 67–73, D-14195 Berlin, Germany.

Contents

Abstract

The antidiuretic hormone arginine-vasopressin (AVP) regulates water reabsorption in renal collecting duct principal cells. Central to its antidiuretic action in mammals is the exocytotic insertion of the water channel aquaporin-2 (AQP2) from intracellular vesicles into the apical membrane of principal cells, an event initiated by an increase in cAMP and activation of protein kinase A. Water is then reabsorbed from the hypotonic urine of the collecting duct. The water channels aquaporin-3 (AQP3) and aquaporin-4 (AQP4), which are constitutively present in the basolateral membrane, allow the exit of water from the cell into the hypertonic interstitium. Withdrawal of the hormone leads to endocytotic retrieval of AQP2 from the cell membrane. The hormone-induced rapid redistribution between the interior of the cell and the cell membrane establishes the basis for the short term regulation of water permeability. In addition water channels (AQP2 and 3) of principal cells are regulated at the level of expression (long term regulation).

This review summarizes the current knowledge on the molecular mechanisms underlying the short and long term regulation of water channels in principal cells. In the first part special emphasis is placed on the proteins involved in short term regulation of AQP2 (SNARE proteins, Rab proteins, cytoskeletal proteins, G proteins, protein kinase A anchoring proteins and endocytotic proteins). In the second part, physiological and pathophysiological stimuli determining the long term regulation are discussed.

1
Introduction: Antidiuresis by Vasopressin

In mammals, the main site of action of a group of nonapeptides termed vasopressins or antidiuretic hormones is the renal collecting duct. Here they induce a rapid increase in the water permeability of the epithelial monolayer, thereby permitting reabsorption of water from the lumen of the collecting duct. As a consequence, urine osmolality increases and urinary output decreases. Vasopressins are, by virtue of their antidiuretic action, the key regulators of water homeostasis. The antidiuretic hormone in man and most mammals is 8-arginine vasopressin.

The molecular targets of vasopressins are heptahelical vasopressin V2 receptors (Birnbaumer et al. 1992) expressed on the basolateral surface of principal cells of the epithelial monolayer lining the renal collecting ducts. Vasopressin V2 receptors are coupled to adenylyl cyclase in a stimulatory fashion *via* the cholera toxin-sensitive G protein G_s. Although other signal pathways have not been rigorously excluded, the hormone-induced rise in intracellular cAMP and subsequent activation of cAMP-dependent protein kinase (PKA) are generally accepted as being responsible for the cellular response (for review see: Breyer and Ando 1994). Inactivating mutations of the vasopressin V2 receptor cause the classical (i.e. X-linked) form of congenital nephrogenic diabetes insipidus (NDI; Rosenthal et al. 1992; for review see: Rosenthal et al. 1996, 1998; Oksche and Rosenthal 1998).

Whereas vasopressin V2 receptors represent the first cellular component of the poorly understood signal transduction cascade activated by vasopressin, the ultimate cellular target of the hormone in kidney principal cells is the water channel protein aquaporin-2 (AQP2) selectively permeable to water (Fushimi et al. 1993; Sasaki et al. 1994). In resting principal cells, AQP2 is localized on intracellular vesicles. According to the "shuttle hypothesis", vasopressin increases water permeability by inducing the exocytotic insertion of AQP2-bearing vesicles, predominantly into the apical plasma membrane of the principal cell. Water is then reabsorbed from the hypotonic urine of the collecting duct. The water channels aquaporin-3 (AQP3) and aquaporin-4 (AQP4), which are constitutively present in the basolateral membrane, allow the exit of water from the cell into the hypertonic interstitium. Originally the renal collecting duct was considered the only AQP2-expressing tissue, but AQP2 expression has recently been reported in endolymphatic sac epithelium of the inner ear (Kumagami et al. 1998).

The water channels AQP2 and AQP3 of principal cells are also regulated at the level of expression by both vasopressin and vasopressin-independent mechanisms (long term regulation of AQP2). Several physiological and pathophysiological conditions are associated with altered expression levels of AQP2. Upregulation may lead to edematic water retention in diseases like congestive heart failure or liver cirrhosis. In contrast, downregulation of AQP2 expression is associated with impaired urinary concentrating ability which may result in polyuria or overt nephrogenic diabetes insipidus (NDI; for review see: Marples et al. 1999).

Inactivating mutations of AQP2 cause autosomal-recessive or autosomal-dominant NDI (Deen et al. 1994; for reviews see: Deen and Knoers 1998a,b; Rosenthal et al. 1996; 1998; Oksche and Rosenthal 1998; Rutishauser and Kopp 1999). Expression of mutants causing autosomal-recessive NDI in *Xenopus laevis* oocytes and in chinese hamster ovary (CHO) cells (G64R, N68S, T126M, A147T, R187C, S216P in oocytes; L22V, T126M, A147T, R187C in CHO cells), showed that they are retained within the endoplasmic reticulum (Deen et al. 1995; Tamaroppoo and Verkman 1998). The recently reported autosomal-dominant form of NDI is caused by a mutation (E258K) which forms heterotetrameres with the wild type protein encoded by the normal allele. The tetramer consisting of normal and mutated proteins is retained within the Golgi apparatus (Mulders et al. 1998; Kamsteeg et al. 1999). Thus in both forms of autosomal NDI, the mutations of AQP2 cause primarily misrouting of the water channel, i.e. the mutated AQP2 is not delivered to the cell surface.

The presence of aquaporin 6 (AQP6) in intracellular vesicles of collecting duct type A intercalating cells has been reported recently (Yasui et al. 1999). Two types of intercalating cells are present in the collecting duct: type A, which secretes H^+, and type B, which secretes HCO_3^-. These cells are responsible, at least in part, for controlling urine pH and acid-base balance (Emmons 1999). Since AQP6 is only present in intracellular vesicles but not in the plasma membrane, this water channel is apparently not involved in water reabsorption from the lumen of the collecting duct.

Since the identification of AQP2, the shuttle hypothesis (see 3) has been confirmed by numerous findings (for review see: Hays 1996). However, the identities of the proteins involved in the signalling cascade downstream of the cAMP/PKA signal and in the shuttle itself remain largely elusive. The purpose of this review is to summarize the current knowledge on the molecular mechanisms underlying the short and long term regulations of water channels in principal cells. In the first part of this review special emphasis is placed on the proteins involved in short term regulation of AQP2 (including SNARE proteins, Rab proteins, cytoskeletal proteins, G proteins, protein kinase A anchoring proteins and endocytotic proteins). In the second part of the review the physiological and pathophysiological stimuli associated with long term regulation will be discussed and the available model systems to study vasopressin-controlled water reabsorption will be compared.

2
Aquaporins Expressed in Principal Cells

An important step towards the understanding of the molecular basis of cellular osmotic water permeabilty was the isolation of an erythrocyte membrane protein, named 28kDa (unglycosylated form) and HMW-28kDa (glycosylated form) by Denker et al. (1988). After cloning of its cDNA (Preston and Agre 1991) and subsequent functional analysis by expression in *Xenopus* oocytes (Preston et al. 1992), the protein was found to be a water channel and hence renamed CHIP28 (Channel-forming Integral Protein of 28 kDa) before it was finally designated AQP1. AQP1 belongs to an ancient protein family, the MIPs (membrane intrinsic proteins), which are abundantly expressed in the lens of the eye. Within the kidney, AQP1 is detectable along the entire proximal tubule and the descending part of the loop of

Henle. This explains the high constitutive water permeability along this nephron segment. Inactivating mutations of AQP1 in man (Preston et al. 1994) did not show significant changes in kidney function, but analysis of AQP1 knock out mice (Ma et al. 1998) showed impaired urinary concentrating ability. Since AQP1 was not detectable along the collecting duct, the existence of another aquaporin appeared likely. The water channel of the collecting duct (WCH–CD), whose existence had been postulated since the 1950s (Koeford-Johnsen and Ussing 1953), was cloned in 1993 (Fushimi et al. 1993). It was later renamed AQP2. Because AQP2 is predominantly localized in the apical membrane of vasopressin-stimulated collecting duct principal cells (for review see: Knepper et al. 1996; Deen and Os 1998), additional water channels were postulated to exist in the basolateral membrane. The first such basolateral water channel present in principal cells to be cloned was AQP3 (Ma et al. 1993a). Functional investigations of AQP3 expressed in oocytes revealed that it was not only permeable to water but also to urea and glycerol (Ishibashi et al. 1994). This water channel was therefore considered to belong to a family of water channels other than the MIP family, more closely related to channels permeable to glycerol, for example to Orf4 in *C. elegans* or the *E. coli* gycerol facilitator GlpF (for review see: Sasaki et al. 1998). Another water channel residing in the basolateral membrane of principal cells was discovered in 1994 (Hasegawa et al. 1994; Jung et al. 1994). In contrast to the mammalian water channels AQP1-3, this water channel, when expressed in *Xenopus* oocytes, was insensitive to the inhibition of water permeability by mercurials, and therefore originally named MIWC (Mercurial-Insensitive Water Channel); it is now designated AQP4. AQP3 is predominantly localized in the lateral membrane compartment, whereas AQP4 seems to be equally distributed between the lateral and basal cell membranes (Terris et al. 1995). The almost exclusive lateral distribution of AQP3 is surprising and difficult to explain, since the basolateral membrane is generally considered as one compartment. However, similar observations were made for the vasopressin V2 receptor, when overexpressed in MDCK cells (Schülein et al. 1998). One may speculate that the lateral localization of AQP3, in combination with its additional permeabilty to urea and glycerol, facilitates transport of these two osmotically active solutes into and out of the lateral intercellular spaces which thus may serve as a reservoir for small osmolytes. Such reservoirs have been reported for osmolytes including sorbitol, betaine and myo-inositol for primary cultured IMCD cells (Ruhfus et al. 1998).

3
The shuttle Hypothesis: From Function to Proteins

In recent years, considerable progress has been made in elucidating the molecular mechanisms underlying the antidiuretic action of vasopressin. A breakthrough was the discovery of aquaporins (Preston et al. 1992), in particular of AQP2 (Fushimi et al. 1993). Prior to this, the effects of vasopressin were mainly monitored using physiological approaches, such as the determination of water permeability changes of epithelia in response to vasopressin.

The first model systems used were amphibian epidermis and urinary bladder (Koeford-Johnsen and Ussing 1953; Hays and Leaf 1962, Bourguet 1976, Stetson 1982, Kachadorian 1985). Both respond to vasopressin with an increase in transepithelial osmotic water permeability. In the apical cell membranes of these epithelia, intramembrane particles (IMPs) were detected by electron microscopy after stimulation with neurohypophysial extracts containing vasopressin (freeze fracture studies; Chevalier et al. 1974). These observations led to the premise that specific structures were responsible for water transport (Hays and Leaf 1962). From mathematical evaluations of physiological experiments it was concluded that vasopressin increases the number of water permeable pores (Hays and Leaf 1962) rather than increasing the size of these pores, as had been previously proposed (Koeford-Johnsen and Ussing 1953). Because of the localization of the IMPs within vesicular structures, an exocytosis-like process seemed to be required for their insertion into cell membranes rather than the regulation of their activity in the cell membrane. Electrophysiological experiments with toad bladders in Ussing-type chambers revealed increases in tissue electrical capacitance after stimulation with vasopressin (Stetson et al. 1982; Palmer et al. 1983), indicating an increase in total cell membrane surface area due to the exocytotic insertion of membranes. IMPs occurred in slightly invaginated membrane patches, resembling clathrin-coated pits (Goldstein 1979). Based on these findings, the membrane shuttle hypothesis was coined to describe the mechanism of vasopressin-regulated water transport in amphibia *via* an exo/endocytotic process (Wade et al. 1981; Handler 1988, Harris et al. 1991, Wade 1994).

Amphibian epidermis or urinary bladder epithelia are useful models for studying the short term regulation of water permeability by vasopressin. However, until now, only one aquaporin (FA-CHIP) has been identified in amphibia (Abrami et al. 1994; Calamita et al. 1994). The highest level of FA-CHIP expression is observed in frog urinary bladder and skin, it is clearly detectable in lung and gall bladder, and low levels are observed in colon and liver tissue; it is, however, not expressed in the kidney. Salt acclimation studies with amphibia demonstrated the induction of FA-CHIP transcription for animals kept in solutions with high osmolality, but there was no evidence for a fast hormonal regulation comparable to that of the mammalian water channel AQP2 (Abrami et al. 1995).

These limitations and the fact that amphibians were not suitable for the study of long term regulation prompted the establishment of mammalian *in-vitro* systems. Burg et al. (1966) reported a method to isolate and study single rabbit nephron fragments in microperfusion experiments. This raised the possibility of comparing results from transepithelial measurements between mammlian and amphibian nephron epithelia under defined conditions. Using this technique, Grantham and Burg (1966) reported an increase in water permeability in the rabbit collecting duct in response to vasopressin. Freeze fracture studies of Harmanci et al. (1978) and Brown and Orci (1983) demonstrated the occurence of an increased number of IMPs in the apical cell membranes of kidney collecting duct cells of water-deprived Wistar rats. Experimental work on isolated perfused mammalian collecting ducts revealed the hydrosmotic effects of vasopressin to be related to a signalling cascade involving the second messenger cAMP (e.g. Grantham and Burg 1966, Snyder et al. 1992, Verkman 1992, Nielsen et al. 1993a). Finally the vasopressin-induced translocation of AQP2-bearing vesicles into the apical plasma membranes of principal cells in rat collecting ducts was shown (Nielsen 1993b). This mechanism establishes the basis of the short term regulation of AQP2.

4
Short Term Regulation: Signals and Proteins Involved in the AQP2 Shuttle

4.1
Cyclic AMP-Triggered Exocytotic Insertion of AQP2 into Apical Cell Membranes and Possible Influence of Intracellular Ca^{2+}

The translocation of AQP2 is an exocytosis-like process involving the recycling of AQP2-bearing vesicles (shuttle hypothesis, see 3). The sequence of events leading to the insertion of AQP2 into cell membranes is comparable to that of the exocytosis of synaptic vesicles, which includes priming, docking and fusion of vesicles and subsequent endocytosis and recycling (for review see: Südhof 1995; Robinson and Martin, 1998). Nevertheless these two exocytotic processes differ fundamentally from each other: whereas a receptor-mediated increase in cAMP is the key trigger for AQP2 translocation, the exocytosis of synaptic vesicles requires an increase in cytosolic Ca^{2+}, which enters the cell *via* voltage-dependent Ca^{2+} channels.

Surprisingly, vasopressin stimulation of principal cells in isolated inner medullary collecting ducts and of cultured collecting duct cells is associated with an increase in intracellular Ca^{2+} (Burnatowska-Hledin and Spielman 1987; Star et al. 1988; Breyer 1991; Champigneulle et al. 1993; Maeda et al. 1993; Ecelbarger et al. 1996). The effect is blocked by the specific vasopressin V2 receptor antagonist $[d(CH_2)_5{}^1,D\text{-}Ile^2, Ile^4, Arg^8]$ (Ecelbarger et al. 1996). Furthermore, vasopressin V1 receptors and oxytocin receptors have been ruled out as being potentially responsible for this Ca^{2+} response (Ecelbarger et al. 1996; Champigneulle et al. 1993). Dual signalling of the vasopressin V2 receptor, i.e. activation of both the G_s/adenylyl cyclase system and the G_q/phospholipase C system has been demonstrated in L cells overexpressing the vasopressin V2 receptor (Zhu et al. 1994). Consequently, a vasopressin V2 receptor-mediated rise of cAMP and intracellular Ca^{2+} was postulated for principal cells (Ecelbarger et al. 1996); but the data on the vasopressin V2 receptor-mediated activation of the G_q/phospholipase C system which results in the production of inositol 1,4,5-trisphosphate (IP_3), the trigger for the release of Ca^{2+} from intracellular stores, are controversial. Teitelbaum (1991) showed a vasopressin-mediated increase in IP_3 production *via* the oxytocin receptor in the rat inner medullary collecting tubule cell line RIMCT, and Gark and Kapturczak (1990) did not find an increase in IP_3 production in isolated rabbit collecting duct cells. In the latter study, an increase in IP_3 production in response to stimulation of the vasopressin V1 receptor was reported. In a more recent report Chou et al. (1998a) showed in rat inner medullary collecting duct cell suspensions that signalling initiated by stimulation of the vasopressin V2 receptor induces neither IP_3 production nor protein kinase C (PKC) activation. Thus the vasopressin V2 receptor-mediated elevation of intracellular Ca^{2+} may occur *via* an IP_3-independent pathway.

One IP_3-independent pathway which may result in the elevation of intracellular Ca^{2+} is initiated by the vasopressin-mediated inhibition of transepithelial Na^+ reabsorption. In cortical collecting ducts vasopressin initially stimulates transepithelial Na^+ absorption resulting in an increase in intracellular Na^+. Thereafter, it causes a sustained inhibition of Na^+ absorption. The transiently elevated Na^+ concentration may activate the Na^+/Ca^{2+} exchanger which could, in turn, enhance the basolateral Ca^{2+} influx. This may contribute to the observed inhibition of Na^+ absorption (Breyer and Ando 1994; Breyer and Fredin 1996).

The role of the increase in intracellular Ca^{2+} in the vasopressin-stimulated increase in osmotic water permeability is not clear. Incubation of

LLC-PK$_1$ cells stably transfected with AQP2 or of primary cultured rat inner medullary collecting duct (IMCD) cells with the PKA inhibitor H89 prevents the exocytotic insertion of AQP2 into the cell membranes (Katsura et al. 1997; Klussmann et al. 1999). Since H89 acts downstream of the G$_s$/adenylyl cyclase system, the vasopressin-induced pathway leading to Ca^{2+} mobilization should not be affected. In addition, forskolin, a direct activator of adenylyl cyclase, induces a strong shuttle response (e.g. Maric et al. 1998). Own data obtained with a membrane-permeable Ca^{2+}-chelating agent (BAPTA-AM) indicate that Ca^{2+} is not required for vasopressin/cAMP-triggered AQP2 translocation (Maric unpublished). These data suggest, that elevation of intracellular cAMP is sufficient for the exocytotic insertion of AQP2 into the apical cell membrane and that elevation of cytosolic Ca^{2+} is not required. However, a modulatory (inhibitory) role of elevated intracellular Ca^{2+} cannot be excluded at the present time (see below).

The vasopressin-mediated increase in water permeability is modulated by several hormones and locally formed factors. In rabbit cortical collecting ducts, basolaterally or luminally administered prostaglandin E$_2$ (PGE$_2$) stimulates osmotic water permeability *via* cAMP production in the absence of vasopressin (Hebert et al. 1993; Sakairi et al. 1995). However, in the presence of vasopressin, basolaterally administered PGE$_2$ inhibits the increase in osmotic water permeability by activation of an inhibitory G protein of the G$_i$ family and inhibition of adenylyl cyclase. Furthermore, it induces elevation of intracellular Ca^{2+} by release from intracellular stores and by influx from the extracellular medium in both rabbit cortical collecting ducts and terminal portions of the rat collecting duct (Hebert et al. 1990; 1991; 1993; Breyer et al. 1990; Breyer 1991). The cellular effects of PGE$_2$ are mediated by G protein coupled EP receptors, of which four distinct subtypes, EP$_1$, EP$_2$, EP$_3$ and EP$_4$, have been cloned (Coleman et al. 1994). EP$_1$ stimulates the G$_q$/phospholipase C system which results in an elevation of intracellular Ca^{2+}; EP$_2$ and EP$_4$ stimulate the G$_s$/adenylyl cyclase system; EP$_3$ activates a G protein of the G$_i$ family and thereby inhibits accumulation of cAMP. *In situ* hybridization experiments have shown the presence of EP$_1$ receptors in human inner and outer medullary collecting ducts and the presence of EP$_3$ and EP$_4$ receptors in human outer medullary collecting ducts (Breyer et al. 1996a; 1996b). The EP receptor pattern expressed by principal cells is unknown, but the inhibitory effect of PGE$_2$ on the vasopressin-stimulated increase in osmotic water permeability is likely to be mediated by EP$_1$ and EP$_3$ receptors (Sakairi et al. 1995; Guan et al. 1998). This is supported by the finding that sulprostone, a EP$_1$/EP$_3$ agonist, inhibits the vasopressin-mediated increases in osmotic water permeability in rabbit cortical collecting duct (Hebert et al. 1993).

Table 4.1. G protein coupled receptors present on inner medullary collecting duct principal cells

Receptor	acting *via*	second messenger
Vasopressin V2	G_s	cAMP↑
Adenosin A1	G_i	cAMP↓
Bradykinin BK2	$G_{q/11}$	Ca^{2+}↑
Muscarinic M $_{1/3?}$	$G_{q/11}$	Ca^{2+}↑
Calcium CaR	$G_{q/11}$ / G_i	Ca^{2+}↑ / cAMP↓
Endothelin ET-B	G_i	Ca^{2+}↑ / cAMP↓
Prostanoid $EP_{1/3}$	$G_{q/11}$ / G_i	Ca^{2+}↑ / cAMP↓

Table 4.1 For explanation see text.

In addition to the G-protein-coupled receptors dicussed above, principal cells express muscarinic receptors, adenosine A1 receptors, bradykinin BK2 receptors and calcium/polycation receptors (Calcium CaR; Table 4.1). Activation of muscarinic and bradykinin BK2 receptors in rat isolated collecting ducts leads to stimulation of phospholipase C presumably *via* the pertussis toxin-insensitive G protein G_q and a subsequent rise in intracellular Ca^{2+} (Snyder et al. 1991; Chou et al. 1998a, Ardaillou et al. 1998). This signalling pathway is consistent with the observed inhibition of the vasopressin-induced increase in osmotic water permeability by the muscarinic agonist carbachol and by bradykinin. Activation of adenosin A1 receptors, causing inhibition of adenylyl cyclase *via* pertussis toxin sensitive G_i proteins, also counteracts the effects of vasopressin on osmotic water permeability (Edwards and Spielman 1994). Elevation of luminal Ca^{2+} inhibits the vaso-pressin-induced increase in osmotic water permeability (for review see: Chattopadhyay et al. 1996), most likely by activation of a calcium/polycation receptor which has recently been shown to be present in the rat terminal collecting duct (Sands et al. 1997). The second messenger system responsible for this effect in principal cells is not clear. In the parathyroid gland a calcium/polycation receptor couples to both the G_q/phospholipase C system (with subsequent elevation of intracellular Ca^{2+}) and to pertussis toxin sensitive G proteins of the G_i family (Garrett et al. 1995).

Endothelin-1 also inhibits the vasopressin-induced increase in osmotic water permeability. The receptor subtype involved is the ET_B receptor, which, by coupling to pertussis toxin-sensitive G_i proteins, activates phospholipase C (with elevation of intracellular Ca^{2+}) and inhibits adenylyl cy-clase (Kohan 1998). Both pathways have been found in collecting ducts. In porcine inner medullary collecting duct cells binding of endothelin-1 to ET_B receptors induces the elevation of intracellular Ca^{2+} (Migas et al. 1993). In rat inner medullary collecting ducts endothelin-1 binding to ET_B receptors inhibits cAMP production (Edwards et al. 1993; Kohan et al. 1993). Therefore inhibition of the vasopressin-induced increase in osmotic water permeability by a number of factors may be at least partially caused by an elevation of intracellular Ca^{2+}. The G protein coupled receptors discussed above are described in Table 4.1.

Atrial natriuretic factor (ANF) binds to a particulate form of guanylyl cyclase (ANF receptor) and thereby stimulates cGMP production. Applied systemically, ANF inhibits not only sodium but also water reabsorption in rat cortical and inner medullary collecting ducts (Nonoguchi et al. 1988; 1989). ANF is also active when applied to the luminal side of the collecting duct (Sonnenberg 1990). However, it is not known whether ANF receptors are present on principal cells.

A potential mechanism, by which Ca^{2+} inhibits the vasopressin-induced increase in osmotic water permeability is suggested by work on Ca^{2+}-sensitive adenylyl cyclases and on phosphodiesterases (Charbades et al. 1996). An elevation of intracellular Ca^{2+} inhibits adenylyl cyclases type V and VI of which type V is present in rat outer medullary collecting duct and type VI is expressed in the entire rat collecting duct (Charbades et al. 1996). PGE_2 reduces vasopressin-stimulated cAMP accumulation in the collecting duct only marginally (up to 12%) in the presence of the non-selective phosphodiesterase (PDE) inhibitor 3-isobutyl-1-methylxanthine (IBMX). In contrast, in the presence of Rolipram (Ro20–1724), a selective inhibitor of cAMP phosphodiesterase type 4 (PDE4), PGE_2 reduces vasopressin-stimulated cAMP accumulation by 43–53% (Charbades et al. 1990; 1996). This result underlines the crucial role of PDE4 activity. This isoform seems to be hyperactive in DI +/+ mice suffering from NDI (Valtin 1992; see 6). The result also suggests that about half of the phosphodiesterase activity has to be accounted for by other phosphodiesterases, of which PDE1, 3 and 5 seem to be present in the rat inner medullary collecting duct (Yamaki et al. 1992; for review see: Dousa 1999). The detection of the Ca^{2+}-dependent PDE1 suggests that the elevation of intracellular Ca^{2+} in response to PGE_2 may activate this PDE isozyme and thereby decreases cAMP levels. This pathway is also conceivable for the inhibitory action of endothelin-1, bradykinin, carbachol and extracellular Ca^{2+} on the vasopressin-stimulated increases in water permeability.

4.2
Comparison of cAMP- and Ca²⁺-Triggered Exocytotic Processes in Various Systems

The classical trigger for exocytotic processes is Ca^{2+}; only occasionally cAMP is the key trigger and in some systems either one of the two second messengers is capable of indepen-dently triggering the exocytotic process. Analysis of the processes in which cAMP is involved may shed some light on the mechanisms of the AQP2 shuttle although the link of cAMP to exocytosis remains elusive in most of the systems discussed below.

An analogous system in which cAMP triggers exocytosis is present in the renin-producing juxtaglomerular cell of the kidney (Kurtz 1989; Hackenthal et al. 1990; Wagner and Kurtz 1998). Stimulation of G_s-coupled β_1-adrenergic receptors expressed on these cells induces an increase in cAMP accumulation (Friis et al. 1999). Prorenin is then secreted into the blood stream and enzymatically cleaved to yield renin. Prorenin release from juxtaglomerular cells is inhibited by binding of angiotensin II to angiotensin AT-1 receptors and subsequent elevation of intracellular Ca^{2+}. Thus renin-producing juxtaglomerular cells and principal cells may be closely related with regard to their regulation of exocytosis.

In other systems both cAMP and Ca^{2+} trigger exocytosis independently from each other. For example, histamine binds to G_s-coupled H_2-receptors on the basolateral membranes of gastric parietal cells. This event leads to activation of adenylyl cyclase and subsequently to an increase in cAMP. Unstimulated parietal cells contain proton pumps (H^+, K^+-ATPases) in the membranes of secretory vesicles. Stimulation of these cells with histamine induces the fusion of secretory vesicles to form canalicular membranes which, in turn, fuse with the apical cell membrane. This cascade drives acid secretion into the stomach lumen (Dransfield et al. 1997). Gastrin independently induces the translocation of the proton pumps into the apical cell membranes through an elevation of intracellular Ca^{2+}. The exocytosis of amylase from the parotid gland is mediated either by cAMP after stimulation of β_1-adrenergic receptors, or, following stimulation of muscarinergic or α_1-adrenergic receptors, by the elevation of intracellular Ca^{2+}. However, the elevation of Ca^{2+} significantly enhances amylase secretion by cAMP. The amount of amylase secreted in response to the elevation of intracellular Ca^{2+} is lower than that released in response to cAMP (Fujita-Yoshigaki et al. 1996; Fujita-Yoshigaki 1998).

Elevation of extracellular glucose is the main trigger for insulin secretion from pancreatic β cells by a mechanism involving closure of ATP-sensitive K^+ channels, depolarization of the cell membrane and a subsequent influx of

Ca^{2+} through voltage-dependent L-type Ca^{2+} channels. The peptide hormone glucagon alone causes insulin release or potentiates insulin secretion by triggering a cAMP/PKA-mediated increase in Ca^{2+} influx through voltage-dependent L-type Ca^{2+} channels (Ämmälä et al. 1993; Wollheim et al. 1996; Lester et al. 1997; Takahashi et al. 1999; Blanpied and Augustine 1999).

Von Willebrand factor is stored in Weibel-Palade bodies in endothelial cells. Exocytosis of the factor occurs in response to thrombin or histamine and other mediators. These agents induce the release of Ca^{2+} from intracellular stores. Alternatively, a receptor-mediated increase in cAMP by iloprost, a synthetic analog of prostacyclin, either induces exocytosis independently of Ca^{2+} or enhances exocytosis evoked by an increase in intracellular Ca^{2+} (Vischer and Wollheim 1998; Vischer et al. 1998).

In many systems, in which Ca^{2+} is sufficient to trigger exocytosis, cAMP plays a modulatory role. For example, in the sensory neurons of *Aplysia*, Ca^{2+} triggers neurotransmitter release which is enhanced by cAMP (for review see: Byrne and Kandel 1996). In pancreatic acinar cells the cholecystokinin (CCK) analogue JMC-180 induces Ca^{2+}-dependently amylase secretion, a process enhanced by the cell membrane-permeable cAMP analog BT_2cAMP-AM (acetoxymethyl ester of dibutyryl cAMP; Vajanaphanich et al. 1995).

The thyroid constitutes a system in which possibly both cAMP and Ca^{2+} are needed for exo- and endocytotic processes. The functional unit of the thyroid is the follicle, comprising a single cell layer lining a luminar space filled with colloid. The follicular cells synthesize thyroglobulin (the storage form of the thyroid hormones T_3 and T_4), transport it in secretory vesicles to their apical cell membranes and secrete it by exocytosis into the colloid. Follicular cells also engulf thyroglobulin by endocytosis. Thyroid stimulating hormone (TSH) -binding to TSH receptors expressed by the follicular cells stimulates both exocytosis and endocytosis of thyroglobulin. Stimulation of the receptor initiates both activation of the G_s/adenylyl cyclase system and the G_q/phospholipase C system. Thus activation of the receptor leads to increases in cAMP and intracellular Ca^{2+} (Allgeier et al. 1993; Wonerow et al. 1998). The exo- and endocytotic processes seem to be stimulated primarily by cAMP; however a role of Ca^{2+} cannot be excluded. The triggers of the exocytotic processes discussed above are summarized in Table 4.2.

Table 4.2. Exocytotic processes classified according to their triggers (cAMP and Ca^{2+})

Exocytotic process	Trigger	Enhancer	Inhibitor
AQP2 insertion into cell membranes	cAMP	?	Ca^{2+} ?
Renin release from juxtaglomerular cells	cAMP	–	Ca^{2+}
HCl release from gastric parietal cells (translocation of H^+/K^+-ATPase)	cAMP or Ca^{2+}	–	–
Amylase release from parotid acinar cells	cAMP or Ca^{2+}	Ca^{2+}	–
Transmitter release from neuronal cells	Ca^{2+}	(cAMP)	–
Insulin release from pancreatic β cells	Ca^{2+} or cAMP	cAMP	–
Von Willebrand factor release from endothelial cells	Ca^{2+} or cAMP	cAMP	–
Amylase secretion from pancreatic acinar cells	Ca^{2+}	cAMP	–
Exo-/endocytosis of thyroglobulin in thyroid follicular epithelial cells	cAMP and Ca^{2+}?	–	–

Table 4.2 For explanation see text.

Another mechanism resembling the exocytotic insertion of AQP2 into cell membranes is the exocytotic insertion of the glucose transporter type 4 (GLUT4) into adipocyte and muscle cell membranes which facilitates glucose uptake (for review see: Olefsky 1999). However major mechanistic differences are likey since GLUT4 translocation is physiologically initiated through substrate phosphorylation by the insulin receptor and subsequent activation of phosphatidyl-inositol 3-kinase and the p21^{ras} pathway.

4.3
Proteins Involved in the Exocytotic Insertion of AQP2 into the Cell Membrane

Over the last couple of years a few proteins have been identified which are potentially involved in AQP2 translocation. In addition, a number of proteins have been identified in principal cells which are assumed to participate in the AQP2 shuttle. However, the complete molecular machinery responsi-

ble for the vasopressin-dependent insertion of AQP2 into cell membranes
and the endocytosis/recycling processes remain largely unknown.

4.3.1
SNARE Proteins

A basic, evolutionarily conserved machinery regulates the fusion of vesicles
with their target membranes (Ferro-Novick and Jahn 1994). One set of key
players involved are membrane proteins referred to as SNAREs (soluble NSF
[N-ethylmaleimide-sensitive fusion protein] attachment protein [SNAP]
receptors); their potential functioning was explained by the SNARE hypothesis originally proposed by Rothman (1994). According to this model,
docking and fusion are mediated by the interaction of SNAREs present on
the vesicle (v-SNAREs) with SNAREs on the target membranes (t-SNAREs).
The SNAREs involved in neuronal exocytosis include the v-SNAREs of the
synaptobrevin family (also referred to as VAMP – vesicle associated membrane protein) and the t-SNAREs syntaxin-1A and SNAP-25 (synaptosome-
associated protein of M_r 25 kDa). These SNAREs assemble spontaneously
into a ternary complex whose core structure has recently been solved
(Sutton et al. 1998). The formation of the complex between proteins of op-
posing membranes is thought to be critical for the fusion of vesicles with
target membrane. In contrast to the original model, recent evidence from
studies on synaptic vesicles and yeast vacuoles indicate that v-t-SNARE
complexes can also assemble on the same membrane (Otto et al. 1997;
Ungermann et al. 1998a). Disassembly of the complexes primes the SNAREs
for fusion (Otto et al. 1997; Ungermann et al. 1998a). This disassembly is
accomplished by soluble proteins, the second set of key players of the fusion
machinery. These are SNAP (soluble NSF attachment protein) proteins (α,
β or γ) which bind to the SNARE complex and NSF (see above), an ATPase
which disrupts the complex by ATP hydrolysis. This modified model may
explain that after priming of the complexes the actual fusion between two
membranes can proceed independently of ATP hydrolysis (Hay and Scheller
1997; Burgoyne and Morgen 1998; Robinson and Martin 1998).

The final phase of vesicle/cell membrane fusion in neuronal cells is controlled by Ca^{2+}, which binds to and activates the Ca^{2+} sensors synaptotagmin
or syncollin, which prevent fusion in the absence of Ca^{2+} ("fusion clamps")
in the exocytosis of secretory vesicles (Brose et al. 1992; Jahn and Südhof
1994; Südhof 1995; Edwardson et al. 1997; Osborne et al. 1999). In yeast
vacuole fusion, the Ca^{2+}-binding protein calmodulin seems to fulfill a post-
docking function, possibly the activation of enzymes that cause bilayer
mixing (Peters and Mayer 1998).

In eukaryotic cells, membrane-bounded compartments exchange their contents and membrane constituents by vesicular trafficking which includes budding of vesicles from donor membranes and fusion with acceptor membranes. Vectoriality in vesicular traffic depends on an accurate and specific membrane recognition and fusion between donor and acceptor. The minimal machinery for membrane fusion includes synaptobrevin, syntaxin and SNAP25 but not the ubiquitously expressed NSF and SNAP proteins (Weber et al. 1998). However, since many relatively distantly related v- and t-SNAREs can pair promiscuously *in vitro* (Fassauer et al. 1999; Yang et al. 1999), the key players of the membrane fusion system described above require additional factors for tethering and docking, to ensure a high degree of specificity (Yang et al. 1999), e.g. small Ras-like GTPases (see 4.3.2). Nevertheless, it is the SNAREs themselves which add to the specificity *in vivo* by virtue of a large number of cell and organelle/vesicle specific isoforms (for review see: Linial 1997).

The analogies between neuronal vesicle exocytosis and AQP2 shuttling led to the search for SNAREs in collecting duct principal cells. We and others identified the v-SNAREs synaptobrevin II and cellubrevin (synaptobrevin III, McMahon et al. 1993) in subcellular fractions enriched for AQP2-bearing vesicles from rat kidney (Liebenhoff and Rosenthal 1995; Jo et al. 1995; Nielsen et al. 1995a). Cellubrevin was also detected in crude preparations of rat inner medullary collecting ducts by Franki et al. (1995a; 1995b). Recently SNAP-23, a homologue of SNAP-25 (Ravichandra et al. 1996; Wang G et al., 1997), was identified on both AQP2-bearing vesicles and at the cell membrane of rat principal cells (Inoue et al. 1998). A potential cognate t-SNARE, syntaxin-4, was identified at the apical cell membrane of rat principal cells (Mandon et al. 1996). Franki et al. (1995a) demonstrated the presence of the soluble NSF and αSNAP proteins in crude preparations of rat kidney inner medulla and amphibian bladder.

The roles of the SNAREs, NSF and SNAP proteins in the AQP2 shuttle remain to be established, although there are hints to their functional involvement. Jo et al. (1995) isolated two types of rat papillary AQP2-containing endosomes, designated light and heavy endosomes according to their buoyant density. Fusion of synaptobrevin-containing light endosomes was inducible by the addition of cytosol and ATP and inhibited in the presence of tetanus toxin which cleaves synaptobrevin II or anti-synaptobrevin antibodies, indicating the involvement of synaptobrevin II in the process. This experiment also suggests that all SNAREs required for fusion are present in the light endosome fraction and catalyze "homotypic" fusion. The physiological significance of the data remains unclear. Light endosomes may be destined to fuse with the apical principal cell membrane.

The functional involvement of synaptobrevin II in cAMP-dependent exocytosis is further supported by the finding that cleavage of synaptobrevin II in rat parotid acinar cells by *Clostridium botulinum* neurotoxin B results in a 52% reduction of amylase release in response to cAMP (Fujita-Yoshigaki et al. 1996). Whether or not SNAP23 and syntaxin-4 are the respective v- and t-SNAREs which participate in the docking and fusion of AQP2-bearing vesicles with the apical cell membrane is questionable because data on the interaction of SNAP-23 and syntaxin-4 with synaptobrevin II are inconclusive. Whereas the human SNAP23 forms SDS-resistant complexes with synaptobrevin IV and VII but not with synaptobrevin II (Yang et al. 1999), the murine SNAP-23 homologue, syndet, interacts with syntaxin-4 and synaptobrevin II (Rea et al. 1998).

In neuronal cells the vesicle-associated synaptophysin binds synaptobrevin and thereby prevents formation of the v-t-SNARE complex. The synaptobrevin-synaptophysin complex has to be disassembled before fusion of vesicles with the cell membrane can occur (Ludger and Galli 1998; Becher et al 1999). Neither synaptophysin nor the Ca^{2+} sensor synaptotagmin I (see above) were detected in AQP2-bearing vesicles isolated from rat collecting duct principal cells (Liebenhoff and Rosenthal 1995). The apparent absence of synaptotagmin I is consistent with a mechanism which does not require Ca^{2+} (see 4.1).

4.3.2
Monomeric GTP-Binding Proteins of the Rab Family

Rab proteins comprise a family of about 40 small GTPases which are involved at different stages of exocytotic and endocytotic membrane trafficking. Like Ras proteins, Rab proteins cycle between an inactive GDP- and an activated GTP-bound state under the control of various regulatory proteins (for review see: Novick and Zerial 1997; Gonzalez and Scheller 1999; Pfeffer 1999).

Most of the Rab proteins are expressed ubiquitously, consistent with their involvement in the membrane traffic common to all cells. However, they differ in their subcellular localization and thereby add to the specificity of vesicle transport and docking and fusion events. For instance Rab6 is associated with intra-Golgi membrane trafficking and Rab3 with vesicles undergoing regulated exocytosis, whereas Rab5 has been detected almost exclusively on endocytotic and recycling vesicles (for review see: Martinez and Goud 1998).

In addition, Rab proteins regulate different stages of the docking and fusion process (transport of vesicles, tethering, docking and fusion; Gonzalez and Scheller 1999). Rab6 may control vesicular transport within the Golgi apparatus by binding Rabkinesin-6, a kinesin-like protein which forms a link between the microtubule-based cytoskeleton and the vesicle-bound Rab-GTP protein (Echard et al. 1998). The yeast homolog of Rab7, Ypt7, has been shown to be required for the tethering of two yeast vacuoles which are destined to fuse (Ungermann et al. 1998b). The Rab5 effector, EEA1 (early endosome antigen 1), may serve a similar function in early endosome tethering (Christoforidis et al. 1999). Schimmöller et al. (1998) proposed that the main function of GTP-bound, vesicle-associated Rab proteins is the recruitment of docking factors to the vesicle. These docking factors might be involved in deprotecting SNAREs (protection is required to prevent random fusion events), thereby facilitating v-t-SNARE pairing and subsequent fusion. Another level of specificity is achieved by cell type-specific expression of some members of the Rab family, e.g. Rab17 (Lutcke et al. 1993) and Rab25 (Goldenring et al. 1993). These two proteins have been detected exclusively in epithelial cells and therefore appear to be associated with vesicle transport specific for epithelial cells. Rab proteins appear not only to be involved in regulating vesicle docking and fusion but also vesicle budding. Overexpression of Rab5 stimulates not only endocytosis but also decreases the number of coated pits on the cell membrane (see 4.4; Bucci et al. 1992).

The Rab3 family consists of Rab3A, B, C and D. Rab3A is enriched on synaptic vesicles and chromaffin granules from bovine adrenal medulla (Fischer von Mollard et al. 1991; Geppert et al. 1994; Slembrouck et al. 1999). Rab3B is localized at the tight junctions of polarized epithelial cells and possibly on synaptic vesicles (Lledo et al. 1993; Weber et al. 1994; Martinez and Goud 1998), Rab3C is found on synaptic vesicles (Fischer von Mollard et al. 1994a) and Rab3D on the zymogen granules of pancreatic acinar cells (Valentijn et al. 1996) and on mast cell secretory vesicles (Tuvim et al. 1999). Thus all Rab3 subtypes are associated with secretory vesicles in cells exhibiting regulated exocytosis. Studies of Rab3A-deficient mice showed that Rab3A is involved in the recruitment of exocytotic vesicles to the cell membrane (Geppert et al. 1994) and, according to more recent data, in the limitation of neurotransmitter release (by controlling the number of synaptic vesicles released; Geppert et al. 1997; for review see: Geppert and Südhof 1998). The mechanism by which Rab3A regulates fusion remains unclear, but its interaction with effector proteins such as raphilin-3A (Shirataki et al. 1993) seems to be important (Gonzalez and Scheller 1999). Rab3A-GTP recruits rabphilin-3A to the vesicle membrane and both proteins dissociate from the vesicle after GTP hydrolysis (Stahl et al. 1995; Geppert et al. 1994; Li et al. 1994). The dissociation of rabphilin-3A from Rab3A-GDP may initiate the interaction of the former with α-actinin, an actin-bundling protein. This interaction permits an enhancement of secretion. Thus Rab3A forms a link between the vesicle and the actin-based cytoskeleton (Kato et al. 1996). Another Rab effector is RIM, a synaptic cell membrane protein. It interacts with vesicle-bound Rab3-GTP, and may be involved in tethering of the vesicle to the cell membrane (Wang Y et al. 1997).

Rab3 proteins, subsequently defined as Rab3A (Klussmann unpublished), coenriched with AQP2 during the purification of AQP2-bearing vesicles from rat renal inner medulla (Liebenhoff and Rosenthal 1995) and were detectable in crude membrane preparations from primary cultured rat inner medullary collecting duct principal cells (Maric et al. 1998). This suggests a possible involvement of Rab3A in the translocation of AQP2 to the cell membrane. These findings together with the observation that SNAREs are present in inner medullary collecting duct cells indicate that the exocytotic insertion of AQP2 into the apical cell membrane is mechanistically related to synaptic vesicle exocytosis. The identification of different Rab proteins in IMCD cells and analysis of their intracellular distribution may allow a differentiation of different pools of AQP2-bearing vesicles, i.e. exo- and endocytotic vesicles.

4.3.3
Cytoskeletal Proteins

The cytoskeleton consists of various functionally distinct components including microtubules, which are composed of α- and β-tubulin, and of actin filaments. The slow-growing (–) end of the microtubule is anchored in the organizing centre, while the fast-growing (+) end projects away from the centre. In nerve cell axons the fast- and slow-growing ends are oriented towards the nerve terminal and the cell body respectively. In polarized epithelial MDCK cells, fast-growing ends project from multiple organizing centres at the apical pole towards the basolateral pole, suggesting that these cytoskeletal filaments might serve as a substrate for the polarized movement of vesicles within the cell (Meads and Schroer 1995). Microtubules are involved in various cellular processes, including the movement of both organelles and vesicles (for review see: Hamm-Alvarez and Sheetz 1998; Bloom and Goldstein 1998). These movements are achieved by microtubule motor proteins which crosslink tubulin to a target organelle and accomplish ATP-dependent transport. One major group of these motor proteins, the dyneins (for review see: Karki and Holzbaur 1999), transport target organelles towards the slow-growing end of the microtubule (retrograde transport). The second group of proteins, the kinesins (for review see: Mandelkow and Hoenger 1999), transport target organelles towards the fast-growing end of the microtubule (anterograde transport). The ability of dynein to mediate transport depends on the presence of soluble factors such as dynactin (Gill et al. 1991; Schroer and Sheetz 1991; for review see: Holleran et al. 1998).

It has long been known that drugs like colchicine, vinblastine, nocodazole and colcemid, which partially depolymerize microtubules, inhibit the vasopressin-induced increase in water permeability of renal collecting ducts (Dousa and Barnes 1974), e.g., nocodazole and colcemid by 65% and 72% respectively (Phillips and Taylor 1989; 1992). Sabolic et al. (1995) showed in immunofluorescence studies of rat principal cells that AQP2-bearing vesicles, normally concentrated beneath the apical cell membrane, become scattered throughout principal cells after colchicine treatment. Cold preservation of kidneys, commonly used during transplantation procedures, induces microtubule disruption, which only partially recovers within 60 min of rewarming. Breton and Brown (1998) showed that cold-induced microtubule disruption inhibits vasopressin-induced AQP2 translocation to the cell membranes in the initial phase of rewarming. Recently Marples et al. (1998) demonstrated the colocalization of dynein and dynactin with AQP2 on immunoisolated AQP2-bearing vesicles, indicating that AQP2-bearing vesicles may indeed undergo transport mediated by dynein and dynactin.

The other component of the cytoskeleton, actin filaments, are also involved in the vasopressin-stimulated increase in osmotic water permeability. Soluble G-actin is assembled into actin filaments (F-actin) under the control of various proteins; this process is reversible. The motor proteins associated with transport along actin microfilaments, such as myosin I and II, belong to the myosin supergene family. Depolymerization of microfilaments by cytochalasin B or dihydrocytochalasin B inhibited concentration-dependently the vasopressin-induced increase in osmotic water permeability in toad bladders by 25–50% suggesting the involvement of microfilaments in the translocation of AQP2 (Taylor et al. 1973; Pearl and Taylor 1983). Surprisingly, vasopressin induces depolymerization of total F-actin in toad bladders by 20–30% (Ding et al. 1991) and of the apical actin in collecting duct principal cells by up to 26% (Simon et al. 1993). In our recently established model of primary cultured rat IMCD cells (Maric et al. 1998), vasopressin and forskolin also induced depolymerization of cortical F-actin (Klussmann unpublished). These findings imply that apical actin is not simply a static barrier to the exocytotic insertion of AQP2 into apical cell membranes but is reorganized after stimulation of principal cells by vasopressin. A likely candidate to direct the reorganization of the actin cytoskeleton is Rab3A (acting via rabphilin-3a) which may be localized on AQP2-bearing vesicles (see 4.3.2).

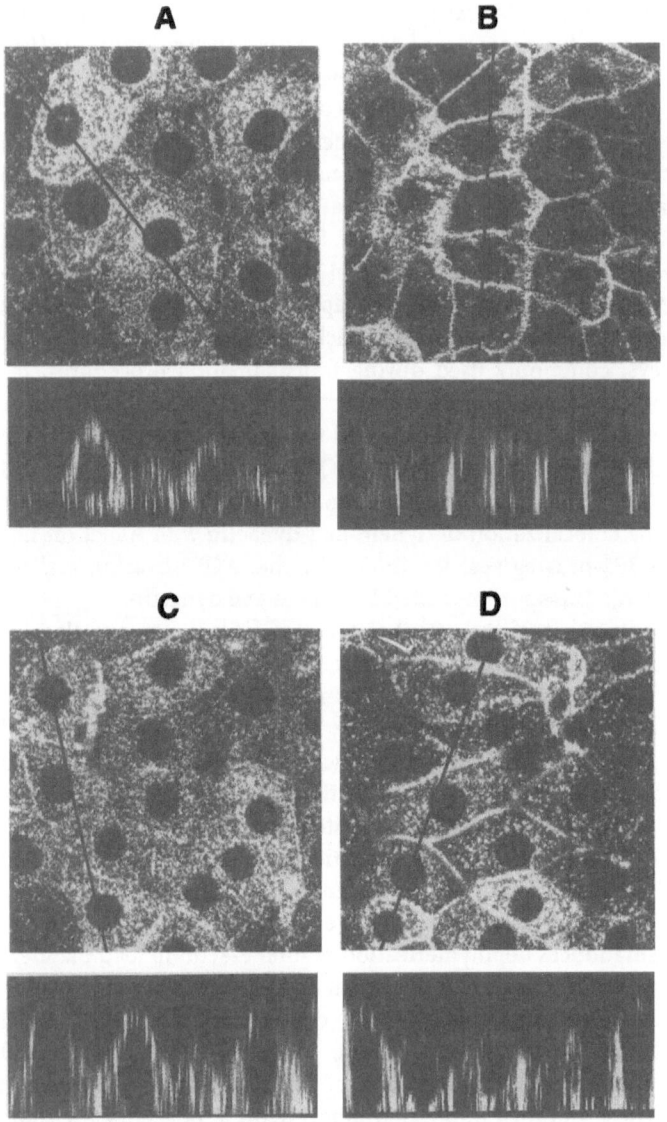

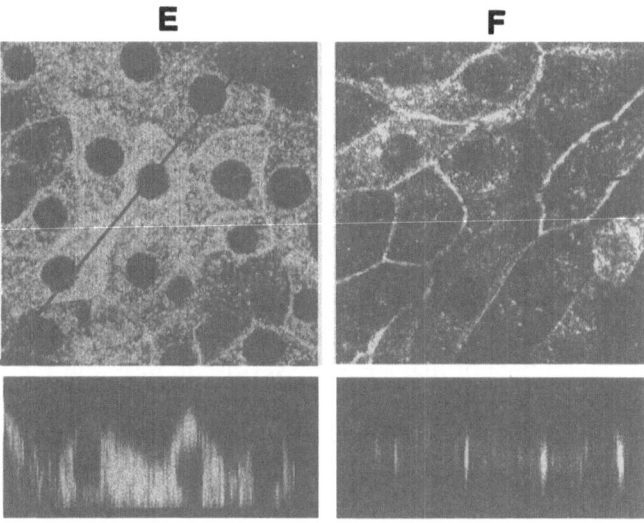

Fig. 4.1. Localization of AQP2 in primary cultured IMCD cells. IMCD cells were incubated with or without synthetic peptides or H89 for 30 min or with *Clostridium difficile* toxin B for 3.5 hours. If indicated, forskolin (100 μM) was added for a further 15 min period. Thereafter cells were fixed, permeabilized, incubated with anti-AQP2 antibodies and secondary cy3-conjugated anti-rabbit antibodies. Immuno-fluorescence was visualized by laser scanning microscopy. The upper part of each panel represents xy scans and the lower part z scans along the black line in the xy image. A: non-stimulated control cells. B: cells stimulated with forskolin. C: forskolin-stimulated cells preincubated with the membrane-permeable peptide S-Ht31 (100 μM) which inhibits binding of PKA to AKAP's (see 4.3.5). D: forskolin-stimulated cells preincubated with S-Ht31-P (100 μM) control peptide. E: forskolin-stimulated cells preincubated with the PKA inhibitor H89 (30 μM). F: non-stimulated cells incubated with *Clostridium difficile* toxin B (400 ng/ml) for 3.5 h. Z-scans in A, C, and E indicate a predominant intracellular localization of AQP2. The z scans in B and D and F demonstrate the presence of AQP2 mainly in the basolateral cell membrane. Adapted from Klussmann et al. (1999)

Another family of monomeric G proteins, the Rho family including Rho, Rac and Cdc42, indirectly controls vesicular traffic by participating through its effectors in the organization of the actin cytoskeleton (for review see: Van Aelst and D'Souza-Schorey 1997; Hall 1998; Aspenström 1999). *Clostridium difficile* toxin B inhibits all small G proteins of the Rho family (Just et al. 1995), while *Clostridium botulinum* C3 toxin inactivates Rho, but not Rac and Cdc42 (Barth et al. 1998; for review see: Aktories 1997; Schmidt and Aktories 1998; Seasholtz et al. 1999). In preliminary experiments we investigated the influence of toxin B on AQP2 translocation in primary cultured rat

IMCD cells (Maric et al. 1998). The toxin induced AQP2 translocation to the cell membrane in the absence of vasopressin, indicating an inhibitory role of Rho proteins, whereby constitutive membrane localization is prevented in unstimulated cells (Fig. 4.1F).

The finding that depolymerization of microtubules only partially inhibits vasopressin-stimulated AQP2 translocation points to two pools of AQP2-bearing vesicles: one pool is dependent upon transport along microtubules, the other (apical) pool is ready to fuse with the cell membrane after hormone stimulation, independently of transport along microtubules. According to this model vasopressin stimulation of collecting duct principal cells induces partial depolymerization of the apical F-actin, allowing release of the apical pool of AQP2-bearing vesicles to fuse with the apical cell membrane; at the same time vasopressin initiates transport of AQP2-bearing vesicles along microtubules towards the apical pole of principal cells. This is consistent with the hypothesis that vasopressin increases the pool of recycling AQP2-bearing vesicles (see 4.4)

The concept of two pools of vesicles is supported by evidence from synaptic vesicle exocytosis in the axon terminal of the presynaptic cell, a "readily releasable" pool exists in the active zone beneath the cell membrane whereas a "reserve pool" is held in the cytoplasm (for review see: von Gersdorff and Matthews 1999). It has also been shown that the exocytosis of dense core vesicles from chromaffin cells occurs in different phases (for review see: Hays et al. 1994; Robinson and Martin 1998).

4.3.4
Heterotrimeric G Proteins

A number of studies have shown that not only small GTP-binding proteins, but also heterotrimeric G proteins are involved in intracellular vesicle trafficking and constitutive and regulated exocytosis (for review see: Helms 1995; Nürnberg and Ahnert-Hilger 1996). The role of heterotrimeric ($\alpha\beta\gamma$) G proteins as signal transducers linking cell surface receptors to various effector systems is well established (for review see: Offermanns and Simon 1996; Vaughan 1998). Like the monomeric GTP-binding proteins, heterotrimeric G proteins cycle between an active (GTP-bound) and inactive (GDP-bound) state. Activated receptors promote the release of GDP from the α subunit, thereby allowing the binding of GTP to the empty subunit. Subsequently, the complex dissociates into a GTP-liganded α subunit and a $\beta\gamma$ heterodimer, both of which regulate by direct interaction the activity of effectors controlling the concentration of intracellular messengers. Hydrolysis of the bound GTP by the GTPase activity intrinsic to the α subunit favours subunit reassociation and thereby terminates G protein action. The newly formed GDP-liganded $\alpha\beta\gamma$ heterotrimer is ready to undergo a new activation/deactivation cycle.

G proteins are not confined to cell membranes but are also present on various intracellular structures including ER and the Golgi apparatus (Helms 1995; 1998), chromaffin granules of PC12 cells, small synaptic vesicles of neurons (Ahnert-Hilger et al. 1994; 1998) and insulin secretory granules of pancreatic β cells (Konrad et al 1995; Kowluru et al. 1996). In MDCK cells, G proteins may play a role in cAMP-stimulated apical endocytosis (Eker et al. 1994). Also in MDCK cells, the G protein G_s enhances vesicular transport from the trans-Golgi network (TGN) to the apical cell membrane; this effect is mimicked by PKA and PKC (Pimplikar and Simons 1994). The splice variant 'extra large' (XL_s) of the G_s α subunit is associated with the trans-Golgi network and occurs in several cell types exhibiting both regulated and constitutive exocytosis (Kehlenbach et al. 1994). A splice variant of the α subunit of G_{i2} (sG_{i2}) associated with the Golgi apparatus of COS-7 cells may also play a role in membrane transport (Montmayeur and Borelli 1994). In MDCK cells a G protein of the G_i family suppress vesicular transport from the TGN to the basolateral membrane (Pimplikar and Simons 1994), and the G protein G_{i3} retards constitutive transport through the Golgi in LLC-PK$_1$ cells, an epithelial cell line derived from porcine proximal kidney tubule (Stow et al. 1991). Dimers of $\beta\gamma$ subunits are involved in the organization of the Golgi apparatus (Yamaguchi et al. 1997; Jamora et al. 1997). In chromaffin cells, activation of G_o proteins associated with the secretory chromaffin granules inhibits exocytosis (Vitale et al. 1993). Gasman et al. (1998) proposed that vesicle-associated G_o controls the peripheral actin cytoskeleton and influences exocytosis by a mechanism involving the activation of RhoA and phosphatidylinositol 4-kinase, both also located on the secretory chromaffin granule. The G protein G_{i3} stimulates exocytosis in mast cells (Aridor et al. 1993; Zussman et al. 1998). Pertussis toxin which prevents activation of G proteins of the G_i/G_o family by ADP-ribosylation of their α subunits inhibits insulin secretion stimulated by mastoparan, a direct activator of G_i (Konrad et al. 1995). Interestingly, the α subunits of G_{i1} regulates the assembly of microtubules by activating the GTPase activity of tubulin (Roychowdhury et al. 1999). In addition, the G protein subunits $G_{\beta1\gamma2}$ bind to and promote the assembly of microtubules (Roychowdhury and Rasenick 1997). These data indicate a possible involvement of G protein α subunits and $\beta\gamma$ complexes in vesicle trafficking along microtubules. Recently it was shown that the introduction of $\beta\gamma$ subunits into permeabilized mast cells enhanced Ca^{2+}-induced secretion and retarded the loss of sensitivity to stimulation usually observed after permeabilization (Pinxteren et al. 1998).

The findings described above led to the investigation of a possible role of heterotrimeric G proteins in the AQP2 shuttle. As a model system, a rabbit cortical collecting duct cell line (CD8 cells) stably transfected with rat AQP2 was used (Valenti et al. 1996). These cells respond to vasopressin or forskolin with a 4–6-fold increase in osmotic water permeability and a translocation of AQP2 to the apical cell membrane. The first evidence for the involvement of G proteins in the AQP2 shuttle came from incubation of CD8 cells with pertussis toxin. The toxin inhibited both the vasopressin- or forskolin-induced increase in osmotic water permeability and the vasopressin- or forskolin-induced translocation of AQP2 to the apical cell membranes (Valenti et al. 1998). *In vitro* ADP-ribosylation of proteins in the AQP2-bearing vesicle preparation demonstrated the presence of two major pertussis toxin-sensitive substrates (40 and 41 kDa). This is consistent with the detection of the α subunits of G_{i2} (40 kDa) and G_{i3} (41 kDa) in Western blot experiments performed with the purified CD8 cell vesicles. In contrast, the α subunits of G_{i1} were hardly detectable in the vesicle fraction of CD8 cells or crude membranes from rabbit kidney but were abundantly present in crude membranes from rat kidney. The α subunits of G_o were detectable neither in vesicles from CD8 cells nor in a crude rat kidney membrane preparation. G protein ß subunits, which are irreversibly associated with γ subunits, were also detected in the vesicle fraction from CD8 cells.

To determine functionally the involvement of the G proteins G_{i2} and G_{i3} in the regulation of AQP2 trafficking, CD8 cells were permeabilized with staphylococcal α toxin (Ahnert-Hilger et al. 1989). The holes formed by the toxin allow the introduction of small molecules and peptides into the interior of the cells. The cells were incubated with synthetic peptides derived from the C termini of α subunits of either $G_{i1/2}$ and G_{i3} ($G_{i1/2}$ and G_{i3} peptide respectively). These peptides contain the cysteine residue modified by pertussis toxin and inhibit the coupling of G proteins to their cognate receptors. Video confocal microscopy showed that the G_{i3} peptide inhibited the cAMP-induced AQP2 translocation concentration-dependently, reaching a maximal inhibitory effect at 50–100 µg/ml. At this concentration, the redistribution of AQP2 appeared to be abolished. In contrast, the $G_{i1/2}$ peptide was inactive at this concentration and showed only a weak inhibitory effect (20–25%) at concentrations up to 200 µg/ml. A control peptide was inactive at a concentration of 200 µg/ml. These data suggest an involvement of G_{i3} in the AQP2 shuttle downstream of the cAMP/PKA signal. Pertussis toxin-sensitive G proteins may therefore stimulate or facilitate insertion of AQP2 into the apical cell membranes. They may either promote early events (i.e. targeting of AQP2-bearing vesicles to the apical cell membrane) or late events (i.e.

docking/fusion of AQP2-bearing vesicles to/with the apical cell membrane; Fig. 7.1).

Immunofluorescence microscopy of CD8 cells revealed a predominantly intracellular localization of the α subunits of G_{i2} and G_{i3}, indicating that G proteins at these locations rather than those associated with the plasma membrane are involved in AQP2 trafficking (Valenti et al. 1998). This is consistent with the association of the α subunits of G_{i3} with zymogen granules from exocrine pancreas where they mediate vesicle swelling, a prerequisite for vesicle fusion with the cell membrane (Jena et al. 1997). In contrast, in mast cells the G protein G_{i3} involved in regulated exocytosis has been located to the cell membrane (Aridor et al. 1993). In LLC-PK$_1$ cells, the α subunits of G_{i3} have been localized on Golgi membranes and shown to play an inhibitory role in the constitutive secretion of heparan sulfate (Stow et al. 1991). Therefore in kidney epithelial cells G_{i3} may play a dual role, i.e. inhibition of constitutive exocytosis and promotion of regulated exocytosis.

In the course of characterizing the signalling from the calcium/polycation receptor (CaR), Sands et al. (1997) demonstrated the presence of α subunits of G_{i1}, G_{i2} and G_{i3} on AQP2-bearing vesicles from rat inner medullary collecting ducts by Western blotting. Our own data are in agreement with theirs regarding the presence of all known α subunits of G_i in the rat kidney preparation. In rabbit kidney (CD8 cells), however, the α subunits of G_{i1} appear to be expressed at low levels. While Sands et al. postulated an inhibitory role of G_i proteins in AQP2 translocation to the cell membrane, i.e. counteracting the effect of vasopressin, through CaR signalling, the data described above demonstrate their involvement in the vasopressin-induced exocytotic insertion of AQP2 into the apical cell membrane.

The mechanism by which G proteins control vesicular transport is unknown, but the present data provide some hints. The finding that the G proteins in fractions enriched for AQP2-bearing vesicles are excellent substrates for pertussis toxin suggests that they are present as $\alpha\beta\gamma$ heterotrimers. G protein heterotrimers are required for interaction with activated cell membrane receptors, and pertussis toxin prevents receptor-mediated but not receptor-independent activation of G proteins (Birnbaumer et al. 1990). The inhibitory effect of pertussis toxin on receptor/G protein coupling, and a plethora of other data (Lambright et al. 1996), show that the C termini of G proteins are important contact sites for receptors. Pertussis toxin and peptides corresponding to the C termini of G_i α subunits inhibit AQP2 translocation to the cell membrane. These findings suggest that, similar to the well known receptor/G protein interactions, a heterotrimeric, pertussis toxin-sensitive G protein is involved in AQP2 translocation to the cell membrane and that the C terminus of its α subunits is required for the interaction with an unknown, possibly upstream, signalling component.

4.3.5
Protein Kinase A Anchoring Proteins (AKAPs)

The binding of AVP to vasopressin V2 receptors (Birnbaumer et al. 1992) on principal cells stimulates cAMP synthesis *via* the G_s/adenylyl cyclase system. Cyclic AMP activates protein kinase A (PKA), which in turn phosphorylates AQP2 at Ser 256 (Fushimi et al. 1997; Lande et al. 1996). The functional consequence of this phosphorylation is not clear: Kuwahara et al. (1995) injected oocytes either with wild-type AQP2 mRNA or with mRNA encoding a mutant AQP2 protein (S256A) which cannot be phosphorylated as was demonstrated by *in vitro* phosphorylation assays. The osmotic water permeability of oocytes injected with wild-type mRNA was 7.7-fold higher than that of water-injected oocytes and increased by an additional 50% after the oocytes were treated with a combination of cAMP and forskolin or with cAMP alone. The osmotic water permeability of oocytes injected with mutant mRNA increased 4.8-fold compared to water-injected oocytes but did not change in response to stimulation with cAMP or forskolin. Lande et al. (1996) found no increase in osmotic water permeability of isolated AQP2-bearing vesicles from rat kidney after treatment with ATP and catalytic subunits of PKA. Elimination of the PKA phosphorylation site of AQP2 (S256A mutant) or incubation with the PKA inhibitor H89 prevents AQP2 translocation in LLC-PK_1 cells stably transfected with the water channel (see 4.1; Katsura et al. 1997). Thus phosphorylation of AQP2 appears to be important for its subcellular localization rather than for its function.

The inactive PKA holoenzyme is a tetramer consisting of a dimer of regulatory subunits (subtypes RI or RII) and two catalytic subunits, each bound to a regulatory subunit. The enzyme is activated by binding of cAMP to the regulatory subunits. Subsequently the catalytic subunits dissociate and phosphorylate their substrates (Taylor et al. 1990; Tasken et al. 1997). Isolated AQP2-bearing vesicles contain PKA activity (Lande et al., 1996), a finding that led to the hypothesis that PKA might be anchored to AQP2-bearing vesicles. The anchoring of PKA to subcellular compartments is mediated by protein kinase A anchoring proteins (AKAPs; Rubin 1994; Faux and Scott 1996; Lester and Scott 1997; Dell'Acqua and Scott 1997; Pawson and Scott 1997; Colledge and Scott 1999). AKAPs bind on one side to a subcellular compartment and on the other to the regulatory PKA subunits (dimer). Most mammalian AKAPs identified so far bind to the RII regulatory subtypes only, with the exception of D-AKAP1 and 2 which bind to both

RI and RII regulatory subtypes (Huang et al. 1997a; 1997b; Banky et al. 1998). The RII-binding domain of AKAPs, which is highly conserved within this protein family, encompasses an amphipathic helix structure. Membrane-permeable peptides (myristylated or stearated) comprising the amphipathic helix of AKAP Ht31 (Carr et al. 1992) effectively compete for PKA-AKAP interaction (Scott 1997; Lester and Scott 1997). These peptides have been shown to relieve cAMP-dependent inhibition of IL-2 transcription (Lester and Scott 1997) and to arrest sperm motility (Vijayaraghavan et al. 1997). In a study using the primary cell culture model of rat IMCD cells (see 6; Maric et al. 1998), the influence of synthetic Ht31-derived peptides coupled to stearic acid (S-Ht31) on vasopressin-mediated AQP2 translocation was tested. Incubation of the cells with S-Ht31 (100 µM) prior to incubation with vasopressin or forskolin inhibited AQP2 translocation to the cell membrane almost completely. The PKA inhibitor H89 (30 µM) abolished vasopressin- or forskolin-induced AQP2 translocation. These data demonstrated that not only the activity of PKA, but also its tethering to subcellular compartments *via* AKAPs are prerequisites for cAMP-dependent AQP2 translocation (Fig. 4.1A–E; Klussmann et al. 1999). This study further showed the presence of different AKAPs in IMCD cell fractions enriched either for AQP2-bearing vesicles or for cell membranes (Klussmann et al. 1999).

PKA compartmentalization is required for various exocytotic processes. PKA anchoring to AKAPs facilitates glucagon-like peptide-1-mediated insulin secretion in primary cultured islets and in rat insulinoma (RINm5F) cells (Lester et al. 1997). Ezrin, a member of the ERM family of AKAPs (Ezrin, Radixin and Moesin; Tsukita et al. 1997; Mangeat et al. 1999; Dransfield et al. 1997) is involved in the translocation of the H^+/K^+ ATPase in parietal cells (Dransfield et al. 1997), a cAMP-dependent exocytotic process analogous to the AQP2 shuttle in kidney principal cells (see 4.2). The expression of ezrin in kidney collecting duct epithelium has been demonstrated (Bretscher et al. 1997) and we identified moesin in principal cells (Klussmann et al. 1998). Ezrin and/or moesin might play a role in AQP2 translocation similar to that of ezrin in gastric acid secretion. Therefore AKAPs may have a general function in cAMP-dependent exocytotic processes (see 4.2).

4.4
Proteins Possibly Involved in Endocytosis

Cells internalize macromolecules such as ligand-bound receptors and other plasma membrane proteins, lipids and small solutes by endocytosis, a process in which the cell membrane invaginates (formation of clathrin-coated pits), captures extracellular components and buds off to form intracellular clathrin-coated vesicles which are transported to the early endosomal compartment. Endocytosed molecules entering the early endosome are either sorted to the late endosomes and lysosomes or are recycled back to the plasma membrane, either directly or through the perinuclear recycling endosomes (for review see: Gruenberg and Maxfield 1995; Schmid 1997; Riezman et al. 1997; Mohrmann and van der Sluijs 1999).

Brown et al. (1988) injected horseradish peroxidase into vasopressin-deficient Brattleboro rats (see 5.1) and measured its internalization by collecting duct cells 15 min later. The internalization of horseradish peroxidase was increased 6-fold by the treatment of the rats with vasopressin for 4 days before injection of horseradish peroxidase (reaching a level comparable to that of normal rats). Nielsen et al. (1993a), in electron microscopic studies of normal rat collecting ducts, showed that the density of coated invaginations resembling clathrin-coated pits increased 5–6-fold per unit area membrane in the presence of vasopressin. In contrast, perfused rabbit cortical collecting ducts show only an insignificant increased internalization of horseradish peroxidase in response to vasopressin compared to untreated controls; however, subsequent removal of vasopressin increased the vesicle uptake about 6-fold (Strange et al. 1988). Katsura et al. (1995) incubated LLC-PK$_1$ cells stably transfected with AQP2 with vasopressin and measured the uptake of FITC-dextran after removal of the hormone. They found a 6-fold increase in internalization of FITC-dextran in cells treated with vasopressin compared to non-treated cells.

The data may indicate that endoytosis is increased in the presence of vasopressin and even further after withdrawal of the hormone. At first glance, it seems difficult to reconcile the stimulatory effect of vasopressin on the removal of AQP2 (endocytosis) from the apical cell membrane with its stimulatory effect on the osmotic water permeability of principal cells, which should depend on the number of AQP2 molecules in the apical cell membrane. This apparent paradox can be solved by assuming that vasopressin increases the pool of AQP2-bearing vesicles undergoing recycling (see 4.3.3). If one further assumes that the "dwell time" of water channels in the apical membrane is not influenced by the hormone, the increase in endocytotic events would simply reflect an increased recycling, a consequence of an increased number of AQP2 molecules in the apical cell membrane. A decrease in the pool of recycling AQP2-bearing vesicles may account for the increase in the uptake of markers from the medium after withdrawal of vasopressin. As described above (see 4.3.5) the mutant S256A of AQP2 is not translocated to the cell membrane (Katsura et al. 1997). Since in LLC-PK$_1$ cells transfected with this mutant, relative endocytosis of FITC-dextran was not stimulated by withdrawal of the hormone this result suggests that the stimulatory effect of vasopressin on endocytosis requires AQP2 to be present in the cell membrane.

The importance of the C terminus for retrieval of AQP2 is further supported by the recent finding of Gustafson et al. (1998) who transfected LLC-PK$_1$ cells with an expression vector encoding a C-terminal fusion of AQP2 with the green fluorescent protein (GFP). The fusion results in a constitutive membrane localization of AQP2, indicating that GFP might mask a signal which is important for the retrieval of AQP2 from the cell membrane (Gustafson et al. 1998). Alternatively the C-terminal GFP might redirect AQP2 from the regulated to the constitutive exocytotic pathway.

The recycling of water channel-containing vesicles in toad bladder and its stimulation by vasopressin was demonstrated by monitoring the endocytotic uptake and subsequent release of horseradish peroxidase into the luminal bath solution. Bladders were treated with vasopressin (for 15 min) which was washed out prior to incubation with horseradish peroxidase (for 10 min). The horseradish peroxidase was then removed and the total uptake determined. After 60 min vasopressin was readded and marker release was measured. At 15 min a release 4-fold higher than in the controls was observed. At 30 min, the release was comparable to that in unstimulated bladders, although water permeability had not declined significantly indicating the recycling of water channels to the cell membranes (Coleman and Wade 1994). Subsequently Katsura et al. (1996) directly demonstrated the recycling of AQP2-bearing vesicles in LLC-PK$_1$ cells stably transfected with AQP2 and treated with cycloheximide, a protein synthesis inhibitor. These studies indicate that at least a portion of AQP2 recycles and is not degraded in the late endosomal/lysosomal compartment.

Several morphological studies of toad bladders and isolated rat collecting ducts support the view that water channel-bearing vesicles pass through distinct endocytotic compartments. Stimulation of the bladders with vasopressin, subsequent removal of the stimulus followed immediately by the addition of markers such as horseradish peroxidase or fluorescent markers to the medium resulted in their appearance in tubular and smaller spherical endosomes within 10 min and in multivesicular bodies within 60 min (Zeidel et al. 1992). Similarly, in studies with isolated rat collecting ducts ferritin or albumin-gold were detected in small endosomes and small multivesicular bodies 5 min and in larger multivesicular bodies 30–60 min after withdrawal of vasopressin (Nielsen et al. 1993a). Studies with toad bladders using pH-sensitive fluid phase markers and fluorescence ratio video microscopy showed that endosomes were not acidified within the first 15 min after endocytosis was induced by vasopressin removal. The transition from neutral to strongly acidic endosomes occured between 30 and 60 min consistent with the appearance of multivesicular bodies (Emma et al. 1994; Nielsen et al. 1993a). For further biochemical analysis subpopulations of endosomes were isolated from toad bladders and rat collecting ducts. In early endosomes from either source the vacuolar H^+-ATPase (responsible for the acidification of endosomes) was not detected (Brown and Sabolic 1993). Acidification is the first step towards the degradative pathway. Thus the lack of acidification of the endosomes is consistent with recycling. The early endosomes contain functional water channels, the apical membrane marker leucine amino peptidase and about 20 other major unidentified proteins (Verkman et al. 1988; Sabolic et al. 1992; Zeidel et al. 1992; Hammond et al. 1993). Harris et al. (1994) showed the presence of leucine amino peptidase and AQP2 in endosomes isolated from rat renal papilla. These endosomes did also not acidify their luminal contents.

Data on the molecular mechanisms leading to endocytosis and recycling of AQP2 are sparse. Electron microscopic studies of isolated collecting ducts from Brattleboro rats showed a parallel appearance of IMPs (see 3) and coated pits resembling clathrin-coated pits in response to a 4 day treatment of the rats with vasopressin (Brown and Orci 1983). The pits were later shown to be coated with clathrin (immunoelectron microscopic studies; Brown et al. 1993). Using collecting ducts of normal Sprague-Dawley rats, Nielsen et al. (1993a) showed a 5–6-fold increase in the density of coated invaginations resembling clathrin-coated pits in the presence of vasopressin. Further evidence for clathrin-mediated endocytosis of AQP2 comes from biochemical studies. Using a procedure (low pH buffer and differential centrifugation) for the isolation of clathrin-coated vesicles from bovine brain, Verkman et al. (1989) isolated vesicles from bovine kidney containing functional water channels. These vesicles resembled clathrin-coated vesicles in electron micrographs. Furthermore, Katsura et al. (1995) showed that K^+ depletion (which inhibits the clathrin pathway) also inhibits the vasopressin-induced increase in FITC-dextran endocytosis in LLC-PK$_1$ cells stably transfected with AQP2, again indicating that retrieval of water channels most likely proceeds through clathrin-coated vesicles. Direct evidence for the clathrin-mediated endocytosis of AQP2 might be forthcoming from studies showing the colocalization of clathrin and AQP2 on vesicles and by demonstrating the functional involvement of proteins of the clathrin pathway in the AQP2 shuttle, in particular of the GTPase dynamin (Sever et al. 1999; Stowell et al. 1999; Takei et al. 1999).

It has been suggested that the ubiquitously expressed protein Rab5 is involved in the formation of clathrin-coated vesicles and formation of early endosomes by fusion of the internalized vesicles (Bucci et al. 1992; Rybin et al. 1996; McLauchlan et al. 1998; Mills et al. 1999). Furthermore Rab5-

interacting soluble factors such as the docking factor early endosomal antigen 1 (EEA1) are also required for endosome fusions (Christoforidis et al. 1999, Sweet 1999). Although present in a subcellular fraction enriched for AQP2-bearing vesicles, Rab5 did not coenrich with AQP2 as did Rab3 (Liebenhoff and Rosenthal 1995). Thus Rab3 and Rab5 may reside on different populations of vesicles which are not separable by the conventionally used Percoll gradient centrifugation step (Sabolic et al. 1992). Whereas Rab3-containing vesicles may represent exocytotic vesicles (see 4.3.2), Rab5-containing vesicles may represent endocytotic vesicles. It is also possible that the distribution of the two Rab proteins overlaps, since the presence of both Rab3A and Rab5 has been demonstrated on synaptic vesicles (Fischer von Mollard et al. 1994b); and the presence of Rab3A has been demonstrated on endosomes isolated from primary cultured bovine chromaffin cells (Slembrouck et al. 1999). Other Rab proteins, such as Rab4, Rab11a and Rab25, are also known to be involved in endocytosis. Transferrin receptor-recycling, for example, is mediated at an early stage by Rab4 and cellubrevin positive vesicles and, at a later stage, by vesicles depleted of Rab4 (Daro et al. 1996). The dominant negative mutant of Rab11 inhibits recycling of internalized transferrin receptors (Ren et al. 1998). Recently Casanova et al. (1999) demonstrated the association of Rab11a and Rab25 with the apical recycling system in MDCK cells. These Rab proteins are also present on immunoisolated proton pump-containing vesicles from gastric parietal cells (see 4.2; Calhoun and Goldenring 1997). Since the proton pumps are translocated to the apical cell membranes in a cAMP-dependent exocytotic process analogous to AQP2 insertion in principal cells, these Rab proteins are prime candidates for governing the recycling of AQP2.

Taken together, the morphological, functional and molecular studies of internalized water channel-bearing vesicles suggest the existence of different populations of endocytotic vesicles, i.e. recycling endosomes and those destined for degradation. Progress in the understanding of AQP2-trafficking will crucially depend on the isolation, functional and molecular characterization of the different classes of AQP2-bearing vesicles.

5
Long-Term Regulation of AQP2

5.1
Regulation of AQP2 Expression by Vasopressin

In addition to its rapid effects leading to an increase in the number of AQP2 molecules in the apical cell membrane (short term regulation), vasopressin also regulates AQP2 expression (long term regulation). Brattleboro rats, (Valtin 1976) suffering from a molecular defect in vasopressin synthesis (Schmale and Richter 1984) were the first mammalian model used for the study of the long term regulation of AQP2. Before specific anti-AQP2 antisera became available, the only possibility of quantitatively investigating the long term regulation of water channels was freeze fracture electron microscopy, allowing the visualization of IMPs, the morphological correlate of aquaporins. IMPs are not detectable in the apical plasma membranes of principal cells from Brattleboro rats. Harmanci et al. (1978) and Brown and Orci (1983) showed that treatment of Brattleboro rats with vasopressin for four days led to the appearence of IMPs. DiGiovanni et al. (1994) conducted long term experiments on Brattleboro rats with implanted osmotic minipumps which constantly released vasopressin over a period of five days. This treatment normalized AQP2 protein expression to the levels observed in Long Evans rats, the unaffected parent strain of Brattleboro rats. Furthermore, perfusion experiments with isolated collecting duct segments from treated and untreated Brattleboro rats confirmed the restored antidiuretic function after long term vasopressin treatment. Increased AQP2 protein levels were also detected in water deprived Wistar rats compared to animals with free access to water (Nielsen et al. 1993b).

In addition, Christensen et al. (1998) showed that AQP2 mRNA levels decreased about 50% in Wistar rats after oral administration of the vasopressin V2 receptor antagonist OPC-31260 for periods of 30 min to 24 hours. Analyses by Yasui et al. (1997) showed that transcription of the human AQP2 requires activation of the transcripton factors CREB and AP-1, both of which then bind to their cognate recognition sites in the AQP2 promoter. In contrast to the human promoter, the rat and mouse AQP2 promoters are devoid of potential AP-1 binding sites (Rai et al. 1997), thus underlining the central role of CREB acting *via* the CRE element for AQP2 expression in these rodents.

5.2
Increased AQP2 Expression under Conditions Associated with Water Retention

The findings described above led to a number of investigations into the expression of AQP2 under both physiological and pathophysiological conditions, specifically those associated with increased water retention during pregnancy, congestive heart failure and liver cirrhosis (for review see: Schrier et al. 1998; Marples et al. 1999). In pregnant rats (Ohara et al. 1998), in rats with experimentally induced congestive heart failure (Xu et al. 1997;

Nielsen et al. 1997) or with liver cirrhosis induced by carbon tetrachloride (CCl_4) (Fujita et al. 1995), a significant increase in AQP2 expression was reported. A more detailed analysis of the animals with congestive heart failure revealed not only an upregulation of AQP2 expression but also a partial misrouting of AQP2 to the basolateral rather than the apical cell membrane of principal cells (Nielsen et al. 1997). Since oral administration of the vasopressin V2 receptor antagonist OPC-31260 reversed the increase in AQP2 expression and normalized symptoms of water retention, it was concluded that elevated levels of circulating vasopressin caused the increase in AQP2 expression.

The key trigger for vasopressin secretion is *hyper*natremia. However, the above mentioned conditions are often associated with *hypo*natremia. Vasopressin secretion is explained in these cases by non-osmotic signalling due to *hypo*volemia because of arterial underfilling (for review see: Schrier et al. 1998). Arterial underfilling has also been observed in liver cirrhosis and pregnancy as a consequence of splanchnic vasodilatation due to increased nitric oxide levels (Niederberger et al. 1995). In heart failure afferent signalling from unloaded high pressure baroreceptors in the atria and the carotids leads to vasopressin secretion (Hadley 1996). In addition, levels of angiotensin-II, a direct stimulator of vasopressin secretion (see 4.1), are elevated in heart failure (Saavedra 1992). Increased vasopressin levels are also known to occur as the syndrome of inappropriate antidiuretic hormone secretion (SIADH), often associated with lung cancer (Bartter and Schwartz 1967). Experimentally induced SIADH results in an increased AQP2 expression (Fujita et al. 1995).

5.3
Decreased AQP2 Expression under Conditions Associated with Decreased Urinary Concentrating Ability

Reports of increased AQP2 expression are still rare, although chronic diseases associated with increased water retention are common. In contrast, diseases associated with decreased or impaired urinary concentrating ability leading to polyuria or the overt manifestation of diabetes insipidus have been reported more frequently and therefore have been more intensively studied with regard to AQP2 expression levels. Three hereditary forms of human diabetes insipidus are recognized in molecular terms: *central* diabetes insipidus caused by impaired vasopressin secretion (Schmale and Richter 1984; Schmale et al. 1993) and two types of NDI (see 1) caused by inactivating mutations of the vasopressin V2 receptor (Rosenthal et al. 1992; for review see: Oksche and Rosenthal 1998; Rosenthal et al. 1998) or the AQP2 genes (Deen et al. 1994; Deen and Knoers 1998a and b; Mulders et al. 1998, Kamsteeg et al. 1999, Rosenthal et al. 1996; 1998; Oksche and Rosenthal 1998; Rutishauser and Kopp 1999). However, these hereditary diseases are very rare (for review see: Robertson 1995).

Acquired forms of polyuria/NDI are often encountered as side effects of drugs, or are associated with metabolic disorders, chronic renal diseases or ureteric obstruction (Robertson 1995). Most cases of acquired NDI are observed during long term lithium treatment of bipolar psychoses; over 50% of lithium-treated patients suffer from acquired NDI (Boton et al. 1987). In most of these patients NDI persists beyond cessation of treatment and may prove irreversible. Among many other effects at the cellular level, lithium has been reported to inhibit the G_s/adenylyl cyclase system and thereby cAMP accumulation in response to stimulation with vasopressin (Yamaki et al. 1991). This effect may contribute to the significant downregulation of AQP2 observed in long term lithium-treated rats (Marples et al. 1995). The irreversible effects of lithium are attributed to general cellular toxicity leading to tubular atrophy and focal sclerosis (Bendz 1983). Another condition, hypokalemia-induced polyuria, is also associated with both decreased cAMP formation and decreased AQP2 expression (Marples et al. 1996). The authors demonstrated the occurence of lysosomes containing granules stained for AQP2. Therefore they postulate an increased degradation of AQP2 under hypokalemic conditions. In contrast to lithium-induced NDI, hypokalemia-induced polyuria is reversible. Normalization of dietary potassium intake leads to normalization of AQP2 expression and the complete restoration of the urinary concentrating ability.

Another severe metabolic disorder associated with acquired NDI is hyperparathyroidism resulting in both hypercalcemia and hypercalciuria (for review see: Robertson 1995; Oksche and Rosenthal 1998). Rats treated with dihydrotachysterol (a vitamin D analog), which increases intestinal Ca^{2+} resorption, develop a 15% increase in plasma calcium levels, a doubling in urinary volume and a marked decrease in AQP2 expression (Earm et al. 1998). The presumably Ca^{2+}-induced downregulation of AQP2 might be caused by activation of luminal Ca^{2+} receptors (CaR) which either might inhibit adenylyl cyclase activity *via* a G protein of the G_i family, or by elevation of intracellular Ca^{2+} (Sands et al. 1997; for review see: Chattopadhyay et al. 1996; see 4.1). The inhibition of vasopressin-regulated water reabsorption in hypercalciuria has been considered as a protective factor, preventing the precipitation of Ca^{2+}-phosphate or Ca^{2+}-oxalate in the lumen of the collecting duct.

Other drugs causing polyuria/NDI have not yet been charcterized with regard to their effect on AQP2 expression and thus their mechanism of action is unknown. The antibiotic demeclocycline causes polyuria possibly by its Mg^{2+}-chelating properties, leading to inhibition of adenylyl cyclase (Mork and Geisler 1995). The barbiturate methohexital, mainly used in experimental general anesthesia, blocks the hydrosmotic response of isolated toad bladder to vasopressin in Ussing-chamber experiments (Stetson et al. 1982). The drug is also thought to inhibit pituitary vasopressin release in the dog (Kasner et al. 1995). Aminoglycoside antibiotics commonly induce NDI due to their general nephrotoxicity. Case reports suggest that the antifungal drug amphotericin B (Höhler et al. 1994) and the antiviral drug cidofovir (Schliefer et al. 1997) also cause NDI.

A decrease in urinary concentrating ability due to reduced AQP2 expression has also been described in the conditions of chronic renal failure and ureteric obstruction. Chronic renal failure was studied using the model of the remnant kidney (removal of 2/3 of one kidney, complete removal of the other kidney; Bonilla-Felix et al. 1992, Teitelbaum and McGuinness 1995 and Kwon et al. 1998). The studies of Teitelbaum and McGuinness (1995) as well as of Kwon et al. (1998) were conducted over a two week period, during which the animals developed renal failure with azotemia, increased serum creatinine levels and polyuria. In addition AQP2 protein levels and vasopressin V2 receptor mRNA levels were decreased (Teitelbaum and McGuinness 1995). The decrease in vasopressin V2 receptor mRNA, possibly caused by locally increased urea concentrations, explains that chronic administration of dDAVP failed to normalize urinary output and AQP2 expression after chronic renal failure had developed (Kwon et al. 1998). A strong decrease in AQP2 levels was also observed after 24 hours of bilateral ureteric obstruction (Frokiaer et al. 1996). The unilateral ureteric obstruction model (Frokiaer et al. 1997) allows differentiation between systemic and renal regulatory mechanisms resulting in decreased AQP2 levels. Whereas in the obstructed kidney AQP2 expression was almost abolished, only a moderate decrease in AQP2 expression was found in the non-obstructed kidney. The authors suggest that high concentrations of urea are responsible for the decrease in AQP2 expression in the obstructed kidney, and that PGE_2 may cause a moderate decrease in the non-obstructed kidney (see 4.1).

5.4
Vasopressin-Independent Regulation of AQP2 Expression

According to the data reviewed so far, long term regulation of AQP2 seems to depend on vasopressin signalling, i.e. an increase in cyclic AMP. There is, however, at least one example of vasopressin-independent downregulation of AQP2, the so called vasopressin escape phenomenon observed in experimental SIADH. Chronic adminstration of dDAVP (implanted minipumps) to rats is associated with an increase in AQP2 levels (Fujita et al. 1995; Ecelbarger et al. 1997). An additional enforced water overload (liquid chow) leads to a strong decrease in AQP2 expression (vasopressin escape phenomenon). The conditions (high dDAVP or vasopressin, water overload) are similar to those found in conditions associated with edematic water retention (pregnancy, congestive heart failure, liver cirrhosis; see 5.2). Yet, SIADH is associated with an decrease and the other conditions with an increase in AQP2 expression. The decrease in AQP2 expression in SIADH is explained by a loss of vasopressin sensitivity of the kidney, experimentally demonstrated by decreased responses (cAMP production, water permeability) of microperfused collecting ducts to vasopressin (Ecelbarger et al. 1998). The authors concluded that at the level of an intact animal vasopressin-independent downregulation of AQP2 exists.

Vasopressin-independent upregulation of AQP2 is observed in rats with lithium-induced NDI. In this form of NDI AQP2 expression is markedly reduced. Marples et al. (1995) found only 10% of AQP2 expression in rats treated with lithium for 35 days compared to untreated control animals. Lithium-induced NDI is largely unresponsive to treatment with vasopressin or dDAVP. Marples et al. (1995) showed that under continued lithium treatment, AQP2 expression is restored to a greater degree by 48 hours thirsting (45% of control) than by a 7 day dDAVP treatment (11% of control). This led to the assumption of vasopressin-independent stimulation of AQP2 expression. Interestingly, urine and plasma osmolality were restored to a lesser extent in thirsted rats than in dDAVP treated rats, and, in contrast to dDAVP, thirsting failed to induce a significant delivery of AQP2 to the apical cell membranes. These data suggest a defect in the AQP2 shuttle to the apical cell membrane. The results of this study should be interpreted with caution since the body weight of the thirsted rats was only about 50% of control rats, indicating extreme dehydration of this group of animals.

6
Model Systems to Study
the Short and Long Term Regulation of AQP2

The first model systems used for the study of vasopressin regulated water transport were of amphibian origin (see 3), but since the identification of AQP2 in human, rat and mouse, especially rats have been used for studying vasopressin-mediated water reabsorption and AQP2 regulation. Wistar, Sprague Dawley, Long Evans and Brattleboro rats have been widely used to investigate either rapid effects of endogenously secreted or exogenously administered vasopressin and to confirm the shuttle hypothesis (see 3). The use of animal models was of particular importance for studying the long term regulation of AQP2 under conditions associated with polyuria/diuresis or edematic water retention (see 5).

In addition, several animal models for diabetes insipidus are availabe. The Brattleboro rat, a model for central diabetes insipidus, has already been introduced (see 3). The DI +/+ mouse is a model or NDI (see 4.1; Valtin 1992). In the latter case, the disease is caused by an abnormally high activity of cAMP phosphodiesterase type 4 (PDE4), and NDI is successfully treated with the PDE4 inhibitor rolipram. AQP1 knock-out mice suffer from diuresis due to a reduced water reabsorption in the proximal tubule (Ma et al. 1998). AQP4 knock-out mice show only a mild defect in urinary concentrating ability (Chou et al. 1998b). AQP2 and AQP3 knock-out mice have not yet been generated. NDI has also been described in other animals including dogs (Luzius et al. 1992) and roosters (Brummermann and Braun 1995). In dogs, NDI is most likely caused by a mutation in the vasopressin V2 receptor (Luzius et al. 1992).

The availability of isolated perfused collecting duct segments from rabbit (Burg et al. 1966) has been a breakthrough for the study of mammalian kidney physiology. After the identification of AQP2, the use of isolated perfused collecting duct segments from rat allowed to correlate osmotic water flow and cell surface expression of AQP2 (Nielsen et al. 1993b). In addition, the model was used to investigate the vasopressin-regulated AQP2 shuttle (see 4). Finally, the model has been useful to investigate functional consequences of the long term regulation of AQP2 (see 5).

In addition, cultured non-epithelial and epithelial cells are used to study the mechanisms of osmotic water permeability changes evoked by vasopressin. The former include *Xenopus laevis* oocytes and chinese hamster ovary (CHO) cells. Oocytes, microinjected with mRNA, are a widely used expression system for functional investigation of proteins. They have been crucial for the functional investigation of AQP2 mutants causing NDI (Deen et al. 1995; Mulders et al. 1998; Kamsteeg et al. 1999; see 1). CHO cells were the first mammalian cells to be stably transfected with water channels (AQP1; Ma et al. 1993b). Recently, this model system was used to study the trafficking and function of naturally occuring mutants of AQP2 (see 1; Tamaroppoo and Verkman 1998).

Polarized systems used for the study of water channels include primary cultured cells and cell lines transfected water channel cDNA. A limitation of primary cultured kidney inner medullary collecting duct cells is the rapid downregulation of AQP2 expression (Furuno et al. 1996). This problem was overcome by modifying the growth conditions (addition of the cyclic AMP analogue dibutyryl cyclic AMP to the medium; Maric et al. 1998). In addition, cell lines overexpressing AQP2 were established. These include: LLC-PK$_1$ cells expressing rat AQP2 C-terminally tagged with c-myc (Katsura et al. 1995), CD8 cells expressing wild-type (untagged) rat AQP2 (Valenti et al. 1996), and WT-10 cells expressing human AQP2 (Deen et al. 1997).

The models described above have preserved key features of principal cells *in-situ*. Most importantly, they exhibit a vasopressin-induced, cAMP-dependent AQP2 shuttle. However they differ in various aspects, in particular in the localization of AQP2: in CD8 and WT-10 cells, stimulatory responses lead to exocytotic insertion of AQP2 into the apical cell membrane. In contrast, in LLC-PK$_1$ cells and primary cultured IMCD cells, a predominant basolateral sorting of AQP2 is observed after cAMP elevation (Fig. 4.1). In the LLC-PK$_1$ cells the C-terminal c-myc tag was discussed to be responsible for the aberrant routing (Katsura et al. 1995). There is no explanation for the basolateral sorting in IMCD cells. There are also differences between CD8 and WT-10 cells: under basal conditions, both cell types show a punctuated intracellular staining when probed with AQP2 antibody. Under stimulated conditions, the staining for AQP2 disappears in CD8 cells (epifluorescence experiments) and is only visible in z-sections calculated from xy-sections obtained by video confocal microscopy. In contrast, in stimulated WT-10 cells apical AQP2 staining can be visualized by conventional epifluorescence microscopy. This finding indicates a more disperse redistribution of AQP2 within the apical cell membrane of CD8 cells than in the other model. Western blot analysis of rat inner medulla suggests that the unglycosylated and glycosylated forms of AQP2 are expressed at approximately equal amounts. However, the glycosylated form of AQP2 is detectable in WT-10 cells only after prolonged exposure of blots (Baumgarten et al. 1998).

AQP2 expression in these transfected cells is driven by a strong viral promoter, precluding experiments on the regulation of AQP2 expression. In contrast, in primary cultured IMCD cells (Maric et al. 1998) AQP2 expression is under the control of the endogenous promoter regulated by the PKA-dependent transcriptionfactor CREB (see 5). Therefore, primary cultured IMCD cells represent a unique *in-vitro* model for central diabetes insipidus and allow studies investigating both long term and short term regulation of AQP2 (Klussmann et al. 1999).

Recently, another cell line expressing AQP2 has been reported by Takacs-Jarrett et al. (1998). This cell line was derived from kidney principal cells of a so called ImmortoMouse carrying a transgene for the SV40 large T antigen. In these cells AQP2 is detectable by Western blotting, and immunofluorescence microscopy (epifluorescence) revealed intracellular staining for AQP2.

However, AQP2 trafficking in response to vasopressin or other agents elevating cAMP has not been reported. It remains to be established whether this model is suited to study the long term regulation of AQP2.

7
Conclusions

The cloning of the vasopressin V2 receptor and of the aquaporins AQP2, 3 and 4 has greatly increased the understanding of the molecular mechanisms of vasopressin-mediated water reabsorption in the collecting duct. In addition, a few proteins have been identified which participate in the AQP2 shuttle.

The current model of the short term regulation of AQP2 (AQP2 shuttle) in principal cells must be refined on the grounds of the findings reviewed in this article. Fig. 7.1 depicts a hypothetical mechanism integrating the proteins that have recently been shown or are assumed to be involved in this process. According to this model, vasopressin binds to the vasopressin V2 receptor on the basolateral surface of principal cells, thereby stimulating adenylyl cyclase *via* the G protein G_s. The hormone-induced rise in intracellular cAMP activates protein kinase A (PKA), whose catalytic subunits subsequently dissociate from the holoenzyme to phosphorylate their substrates, one of which is AQP2. This phosphorylation leads to the exocytotic insertion of AQP2 primarily into the apical cell membrane. To trigger the exocytotic process, PKA must be anchored to subcellular compartments *via* AKAPs. These compartments have not yet been defined, but AKAPs appear to be present on AQP2-bearing vesicles and the cell membrane, and some may be associated with the cytoskeleton. Downstream of the cAMP/PKA signal a heterotrimeric G protein of the G_i family, probably G_{i3}, is involved in the exocytotic insertion of AQP2 into the cell membrane, though its exact function, in particular its upstream and downstream partners, has not yet been defined. Rab3A, possibly residing on AQP2-bearing vesicles, may predestine them for exocytotic insertion into the cell membrane. The docking/fusion of the vesicles with the target membrane might involve v-and t-SNAREs which reside either on the vesicle (synaptobrevin II), on the cell membrane (syntaxin 4) or on both (SNAP23). AQP2-bearing vesicles directed to fusion with the cell membrane are transported along microtubule filaments, possibly using dynein as the motor protein. The cortical actin cytoskeleton seems to represent a barrier for AQP2-bearing vesicles and must be reorganized prior to their fusion with the cell membrane.

Internalization of AQP2 may be initiated by the formation of clathrin-coated pits followed by the formation of clathrin-coated vesicles. Rab5 may

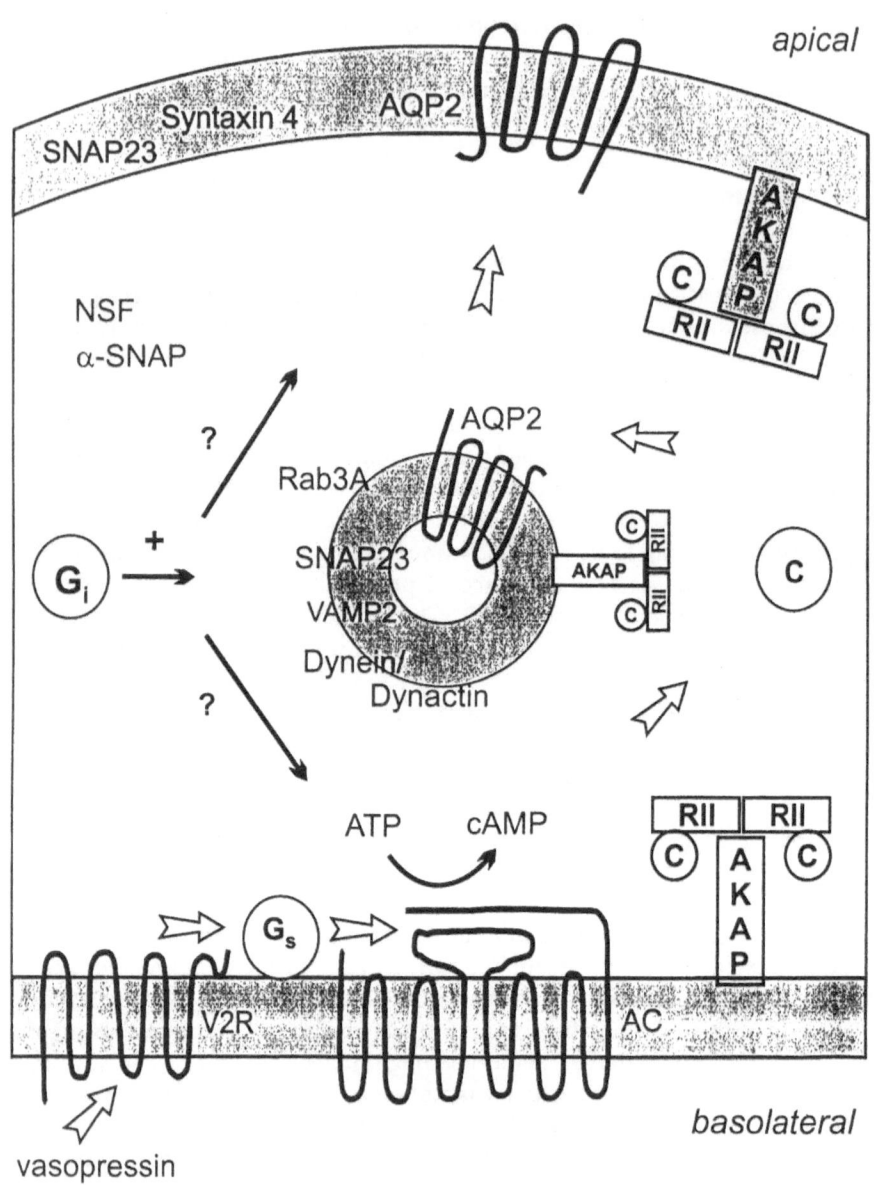

regulate the formation of endocytotic vesicles which enter the early endosomal compartment from where they may be recycled back to the cell membrane or ultimately fuse with lysosomes.

Despite considerable progress, many questions remain. The intriguing question, how the vasopressin-induced increase in cAMP and the subsequent steps (activation of PKA and phosphorylation of AQP2 at Serin 256) eventually leads to the exocytotic insertion of AQP2 molecules into the cell membrane, still awaits an answer. The current data are consistent with the hypothesis that these early events primarily lead to an increase in the recycling pool of AQP2. In addition, the precise roles of the proteins shown to be involved in the AQP2 shuttle remain to be determined, and further components required for this process have to be identified.

The mechanism by which hormones and locally produced factors inhibit the vasopressin-induced increase in osmotic water permeability is not yet understood. Inhibition of vasopressin's action by receptor agonists such as PGE_2, endothelin-1, bradykinin or carbachol is associated with an elevation

---

Fig. 7.1. A model for the vasopressin/cAMP-triggered exocytotic insertion of AQP2 into the apical cell membranes of a renal collecting duct principal cell. All proteins with known functional relevance to AQP2 translocation are depicted as symbols; those which have been shown to be associated with either the AQP2-bearing vesicles or the cell membranes but whose function has not yet been delineated are indicated by name. The molecular targets of vasopressin are heptahelical vasopressin V2 receptors (V2R; Birnbaumer et al. 1992) expressed on the basolateral surface of principal cells. V2 receptors are coupled to adenylyl cyclase (AC) in a stimulatory fashion *via* the cholera toxin-sensitive G protein G_s. The hormone induces a rise in intracellular cAMP which activates protein kinase A (PKA) which is anchored to several subcellular compartments by protein kinase A anchoring proteins (AKAPs; Klussmann et al. 1999). The inactive PKA holoenzyme is a tetramer consisting of a dimer of regulatory subunits (RII) and two catalytic subunits (C), each of which is bound to a regulatory subunit. Binding of cAMP to the regulatory subunits leads to dissociation and activation of catalytic subunits, which catalyse phosphorylation of their substrates, including AQP2. Furthermore a G protein of the G_i family, most likely G_{i3}, is required downstream of the cAMP/PKA signal (Valenti et al. 1998). The mechanisms underlying the docking and fusion of AQP2-bearing vesicles predominantly with the apical cell membrane are not known. The detection of the membrane-associated Rab3A (Liebenhoff and Rosenthal 1995; unpublished results), synaptobrevin II, here referred to as VAMP2 (Liebenhoff and Rosenthal 1995; Nielsen et al. 1995a; Jo et al. 1995), syntaxin 4 (Mandon et al. 1996), SNAP23 (Inoue et al. 1998) and of the soluble proteins NSF and α–SNAP (Franki et al. 1995a) suggests that these proteins are involved in docking and fusion of AQP2-bearing vesicles. Intracellular transport of AQP2-bearing vesicles may be achieved by trafficking along microtubule filaments using dynein as the motor protein in conjunction with dynactin (Marples et al. 1998)

of intracellular Ca^{2+}. However, whether and how an increase in intracellular Ca^{2+} inhibits the AQP2 shuttle is not known.

Elucidation of the molecular basis of the AQP2 shuttle will have an impact on the understanding of cAMP-dependent exocytotic processes in other cell types, e.g. renin release from juxtaglomerular cells, HCl release from gastric parietal cells, or amylase release from parotid acinar cells. The finding that SNAREs known to participate in Ca^{2+}-triggered exocytosis in secretory cells (e.g., neuronal and neuroendorine cells) are also present in principal cells suggests that the Ca^{2+}-triggered exocytosis and the cAMP-induced AQP2 shuttle are mechanistically similar. This similarity may be confined to late steps (docking and fusion), whereas the early steps (recruitment of vesicles and targeting to the „active zone" within the cells) are likely to differ fundamentally, since they are triggered by different signals (Ca^{2+} *versus* cAMP).

Increased understanding of the molecular basis of the vasopressin-induced AQP2 shuttle should also have major diagnostic and therapeutic implications for the treatment of pathological states associated with water retention or a loss of body water. The discovery that mutations in the vasopressin V2 receptor gene or the AQP2 gene are responsible for congenital NDI already allows an early diagnosis of the disease, which is of great clinical significance (for review see: Oksche and Rosenthal 1998). The identification of proteins involved in the AQP2 shuttle should pave the way to the exploration of new therapeutic strategies. Intracellular drug targets are of particular interest since they may allow activation of the cascade without the involvement of the vasopressin V2 receptor. This is a major advantage for patients with receptor defects, i.e. those with X-linked NDI.

Most of the mutans of the vasopressin V2 receptor and AQP2, causing X-linked or autosomal NDI respectively, are misfolded and retained in the ER, similar to many other membrane protein mutants causing disease (Aridor and Balch 1999). Tamarappoo and Verkman (1998) showed that intracellularly retained AQP2 mutants are rescued (i.e. appear on the cell surface) following the treatment of CHO cells with chemical chaperones like trimethylamine oxide (TMAO). Gene therapy is another option for the treatment of congenital forms of NDI and other tubular defects caused by mutations in single genes, since genes can be easily targeted to the tubular system of the kidney (retograde transfer of genes; Moullier et al. 1994). This however requires that the principle obstacles preventing the successful implementation of gene therapy will be overcome in the future.

Various models are available to study new therapies and drugs for the treatment of diseases or symptoms caused by disturbances of water reabsorption. Brattleboro rats and DI +/+ mice represent animal models for central and NDI respectively (Valtin 1976; 1992). Several cell culture models are available to study the effect of drugs on osmotic water permeability and AQP2 shuttle *in vitro*. These include cell lines derived from LLC-PK$_1$ cells (Katsura et al. 1995), CD8 cells (Valenti et al. 1996), MDCK cells (Deen et al. 1997) and CHO cells (Tamaroppoo and Verkman 1998), all of which overexpress AQP2. In addition, primary cultured IMCD cells from rat kidney, which have maintained AQP2 expression (Maric et al. 1998) are available. The latter model is not only suitable for the study of the short term effect of drugs (AQP2 shuttle), but also allow the study of long term effects (expression of AQP2 and of other proteins involved in the shuttle). Furthermore, a cell line derived from the so called ImmortoMouse has been established, although it is not known whether these cells show a vasopressin-sensitive AQP2 shuttle (Takacs-Jarrett et al. 1998). New insights in the molecular basis of water reabsorption and the availability of various animal and cell culture models provide an excellent basis for the development of new forms of treatment for patients with symptoms caused by abnormalities in water reabsorption.

Acknowledgements. We thank John Dickson for critical reading of the manuscript. This review includes work of our colleagues M. Beyermann, A. Geelhaar, E. Krause, R. Storm, B. Wiesner, and G. Valenti. Own work reported herein was supported by grant Ro 597/6-2 from the Deutsche Forschungsgemeinschaft to W. R.

References

Abrami L, Simon M, Rousselet G, Berthonaud V, Buhler JM, Ripoche P (1994) Sequence and functional expression of an amphibian water channel, FA-CHIP: a new member of the MIP family. Biochim Biophys Acta 1192:147–151

Abrami L, Capurro C, Ibarra C, Parisi M, Buhler JM, Ripoche P (1995) Distribution of mRNA encoding the FA-CHIP water channel in amphibian tissues:effects of salt adaptation. J Membrane Biol 143:199–205

Ahnert-Hilger G, Mach W, Fohr KJ, Gratzl M (1989) Poration by alpha-toxin and streptolysin O: an approach to analyze intracellular processes. Methods Cell Biol 31:63–90

Ahnert-Hilger G, Schäfer T, Spicher K, Grund C, Schultz G, Wiedenmann B (1994) Detection of G protein heterotrimers on large dense core and small synaptic vesicles of neuroendocrine and neuronal cells. Eur J Cell Biol 65:26–28

Ahnert-Hilger G, Nürnberg B, Exner T, Schäfer T, Jahn R (1998) The heterotrimeric G protein G_{o2} regulates catecholamine uptake by secretory vesicles. EMBO J 17:406–413

Aktories K (1997) Rho proteins:targets for bacterial toxins. Trends Microbiol 5:282–288

Allgeier A, Offermanns S, Van Sande J, Spicher K, Schultz G, Dumont JE (1993) The human thyrotropin receptor activates G-proteins Gs and Gq/11. J Biol Chem 269:13733–13735

Ämmälä C, Ashcroft FM, Rorsman P (1993) Calcium-independent potentiation of insulin release by cyclic AMP in single β cells. Nature 363:356–358

Ardaillou N, Placier S, Zhao J, Baudouin B, Ardaillou R (1998) Characterization or B2-bradykinin receptors in rabbit principal cells of the collecting duct. Exp Nephrol 6:534–541

Aridor M, Rajmilevich G, Beaven MA, Sagi-Eisenberg R (1993) Activation of exocytosis by the heterotrimeric G protein G_{i3}. Science 262:1569–1572

Aridor M, Balch WE (1999) Integration of cytoplasmic reticulum signaling in health and disease. Nat Med 5:745–751

Aspenström P (1999) The Rho GTPases have multiple effects on the actin cytoskeleton. Exp Cell Res 246:20–25

Banky P, Huang LJ, Taylor SS (1998) Dimerization/docking domain of the type Ialpha regulatory subunit of cAMP-dependent protein kinase. Requirements for dimerization and docking are distinct but overlapping. J Biol Chem 273:35048–35055

Barth H, Hofmann F, Olenik C, Just I, Aktories K (1998) The N-terminal part of the enzyme component (C2I) of the binary Clostridium botulinum C2 toxin interacts with the binding component C2II and functions as a carrier system for a Rho ADP-ribosylating C3-like fusion toxin. Infect Immun 66:1364–9

Bartter FC, Schwartz WB (1967) The syndrome of inappropriate secretion of antidiuretic hormone. Am J Med 42:790–806

Baumgarten R, van den Pol MHJ, Wetzels JFM, van Os CH, Deen PMT (1998) Gycosylation is not essential for vasopressin dependent routing of aquaporin-2 in transfected Madin-Darby canine kidney cells. J Am Soc Nephrol 9:1553–1559

Becher A, Drenckhahn A, Pahner I, Margittai M, Jahn R (1999) The synaptophysin-synaptobrevin complex:a hallmark of synaptic vesicle maturation. J Neurosci 19:1922–1931

Bendz H (1983) Kidney function in lithium treated patients. A Literature survey. Acta Psychatr Scand 68:303–324

Birnbaumer L, Abramowitz J, Brown AM (1990) Receptor-effector coupling by G proteins. Biochim Biophys Acta 1031:163–224

Birnbaumer M, Seibold A, Gilbert S, Ishido M, Barberis B, Antaramian A, Brabet P, Rosenthal, W (1992). Molecular cloning of the receptor for human antidiuretic hormone. Nature 357:333–335

Blanpied TA, Augustine GJ (1999) Commentary:Protein kinase A takes center stage in ATP-dependent insulin secretion. Proc Natl Acad Sci USA 96:329–331

Bloom GS, Goldstein LSB (1998) Cruising along microtubule highways:how membranes move through the secretory pathway. J Cell Biol 140:1277–1280

Bonilla-Felix M, Hamm LL, Herndon J, Vehaskari VM (1992) Response of cortical collecting ducts from remnant kidneys to arginine vasopressin. Kidney Int 41:1150–1154

Boton R, Gaviria M, Battle DC (1987) Prevalence, pathogenesis and treatment of renal dysfunction associated with chronic lithium therapy. Am J Kidney Dis 10:329–345

Bourguet J, Chevalier J, Hugon JS (1976) Alterations in membrane associated particle dirstribution during antidiuretic challenge in frog urinary epithelium. Biophys J 16:627–639

Bretscher A, Reczek D, Berryman M (1997). Ezrin:a protein requiring conformational activation to link microfilaments to the plasma membrane in the assembly of cell surface structures. J Cell Sci 110:3011-3018

Breton S, Brown D (1998) Cold-induced microtubule disruption and relocalization of membrane proteins in kidney epithelial cells. J Am Soc Nephrol 9:155-166

Breyer MD, Jacobson HR, Hebert RL (1990) Cellular mechanisms of prostaglandin E2 and vasopressin interactions in the collecting duct. Kidney Int 38:618-624.

Breyer MD (1991) Regulation of water and salt transport in collecting duct through calcium-dependent signaling mechanisms. Am J Physiol 260:F1-F11

Breyer MD, Ando Y (1994) Hormonal signaling and regulation of salt and water transport in the collecting duct. Annu Rev Physiol 56:711-739

Breyer MD, Fredin D (1996) Effect of vasopressin on intracellular Na^+ concentration in cortical collecting duct. Kidney Int Suppl 57:S57-S61

Breyer MD, Jacobson HR, Breyer RM (1996a) Functional and molecular aspects of renal prostaglandin receptors. J Am Soc Nephrol 7-17

Breyer MD, Davies L, Jacobson HR, Breyer RM (1996b) Differential localization of prostaglandin E receptor subtypes in human kidney. Am J Physiol F912-F918

Brose N, Petrenko AG, Südhof TC, Jahn R (1992) Synaptotagmin: a calcium sensor on the synaptic vesicle surface. Science 256:1021-1025

Brown D, Orci L (1983) Vasopressin stimulates the formation of coated pits in rat kidney collecting ducts. Nature 302:253-255

Brown D, Weyer P, Orci L (1988) Vasopressin stimulates endocytosis in kidney collecting duct epithelial cells. Eur J Cell Biol 46:336-340

Brown D, Sabolic I (1993) Endosomal pathways for water channel and proton pump recycling in kidney epithelial cells. J Cell Sci 17:49-59

Brummermann M, Braun EJ (1995) Renal response of roosters with diabetes insipidus to infusion of arginine vasotocin. Am J Physiol 269:R57-R63

Bucci C, Parton RG, Mather IH, Stunnenberg H, Simons K, Hoflack B, Zerial M (1992) The small GTPase rab5 functions as a regulatory factor in the early endocytic pathway. Cell 70:715-728

Burg M, Grantham J, Abranow M, Orloff J (1966) Preparation and study of fragments of single rabbit nephrons. Am J Physiol 210:1293-1298

Burgoyne RD, Morgan A (1998) Analysis of regulated exocytosis in adrenal chromaffin cells:insights into NSF/SNAP/SNARE function. BioEssays 20:328-335

Burnatowska-Hledin MA, Spielman WS (1987) Vasopressin increases cytosolic free calcium in LLC-PK_1 cells through a V_1-receptor. Am J Physiol 253:F328-F332

Byrne JH and Kandel ER (1996) Presynaptic facilitation revisited:state and time dependence. J Neurosci 16:425-435

Calamita G, Mola MG, Svelto M (1994) Presence in frog urinary bladder of proteins immunologically related to the aquaporin-CHIP. Eur J Cell Biol 64:222-228

Calhoun BC, Goldenring JR (1997) Two Rab proteins, vesicle-associated membrane protein 2 (VAMP-2) and secretory carrier membrane proteins (SCAMPs), are present on immunoisolated parietal cell tubulovesicles. Biochem J 325:559-564

Carr DW, Hausken ZE, Fraser IDC, Stofko-Hahn RE, Scott JD (1992) Association of the type II cAMP-dependent protein kinase with a human thyroid RII-anchoring protein. J Biol Chem 267:13376-13382

Casanova JE, Wang X, Kumar R, Bhartur SG, Navarre J, Woodrum JE, Altschuler Y, Ray GS, Goldenring JR (1999) Association of Rab25 and Rab11a with the apical recycling system of polarized Madin-Darby canine kidney cells. Mol Biol Cell 10:47-61.

80   E. Klussmann et al.

Chabardes D, Montegut M, Zhou Y, Siaume-Perez S (1990) Two mechanisms of inhibition by prostaglandin E2 of hormone-dependent cell cAMP in the rat collecting tubule. Mol Cell Endocrinol 73:111–21

Chabades D, Firsov D, Aarab L, Clabecq A, Bellanger A-C, Siaume-Perez S, Elalouf J-M (1996) Localization of mRNAs encoding Ca^{2+}-inhibitable adenylyl cyclases along the renal tubule. J Biol Chem 271:19264–19271

Champigneulle A, Siga E, Vassent G, Imbert-Teboul M (1993) V_2-like vasopressin receptor mobilizes intracellular Ca^{2+} in rat medullary collecting tubules. Am J Physiol 265:F35–F45

Chattopadhyay N, Mithal A, Brown EM (1996) The calcium-sensing receptor: a window into the physiology and pathophysiology of mineral ion metabolism. Endocr Rev 17:289–307

Chevalier J, Bourguet J, Hugon JS (1974) Membrane associated particles:Distribution in frog urinary bladder epithelium at rest and after oxytocin treatment. Cell Tiss Res 152:129–140

Chou C-L, Rapko SI, Knepper MA (1998a) Phosphoinositide signaling in rat inner medullary collecting duct. Am J Physiol 274:F564–F572

Chou C-L, Ma T, Yang B, Knepper MA, Verkman AS (1998b) Fourfold reduction of water permeability in inner medullary collecting duct of aquaporin-4 knock out mice. Am J Physiol 274:F549–F554

Christensen BM, Marples D, Jensen UB, Frokiaer J, Sheikh-Hamad D, Knepper M, Nielsen S (1998) Acute effects of vasopressin V2-receptor antagonist on kidney AQP2 expression and subcellular distribution. Am J Physiol 275:F285–F297.

Christoforidis S, McBride HM, Burgoyne RD, Zerial M (1999) The Rab5 effector EEA1 is a core component of endosome docking. Nature 397:621–5

Coleman RA, Wade JB (1994) ADH-induced recycling of fluid-phase marker from endosomes to the mucosal surface in toad bladder. Am J Physiol 267:C32–38.

Coleman RA, Smith WL, Narumiya S (1994) International union of pharmacology classification of prostanoid receptors:properties, distribution, and structure of the receptors and their subtypes. VIII Pharmacol Rev 46:205–229

Colledge M, Scott JD (1999) AKAPs:from structure to function. Trends Cell Biol 9:216–221

Daro E, van der Sluijs P, Galli T, Mellman I (1996) Rab4 and cellubrevin define different endosome populations on the pathway of transferrin receptor recycling. Proc Natl Acad Sci USA 93:9559–9564

Deen PMT, Verdijk MAJ, Knoers NVAM, Wieringa B, Monnens LAH, van Os CH, van Oost BA (1994) Requirement of human renal water channel aquaporin-2 for vasopressin-dependent concentration of urine. Science 264:92–95

Deen PM, Croes H, van Aubel RA, Ginsel LA, van Os CH (1995) Water channels encoded by mutant aquaporin-2 genes in nephrogenic diabetes insipidus are impaired in their cellular routing. J Clin Invest 95:2291–2296

Deen PMT, Rijss JPL, Mulders SM, Errington RJ, Baal JV, van Os CH (1997) Aquaporin-2 transfection of Madin-Darby canine kidney cells reconstitutes vasopressin-regulated transcellular osmotic water transport. J Am Soc Nephrol 8:1493–1501

Deen PMT, Knoers NVAM (1998a) Physiology and pathophysiology of the aquaporin-2 water channel. Curr Opin Nephrol Hypertens 7:37–42

Deen PMT, van Os CH (1998) Epithelial aquaporins. Curr Opin Cell Biol 10:435–442

Deen PM, Knoers NV (1998b) Vasopressin type-2 receptor and aquaporin-2 water channel mutants in nephrogenic diabetes insipidus. Am J Med Sci 316:300–9

Dell'Acqua ML, Scott JD (1997) Protein kinase A anchoring. J Biol Chem 272:12881–12884

Denker BM, Smith BL, Kuhajda FP, Agre P (1988) Identification, purification, and partial characterization of a novel Mr 28,000 integral membrane protein from erythrocytes and renal tubules. J Biol Chem 263:15634–15642

DiGiovanni SR, Nielsen S, Christensen EI, Knepper MA (1994) Regulation of collecting duct water channel expression by vasopressin in Brattleboro rat. Proc Natl Acad Sci 91:8984–8988

Ding G, Franki N, Condeelis J, Hays RM (1991) Vasopressin depolymerizes F-actin in toad bladder epithelial cells. Am J Physiol 260:C9–C16

Dousa TP, Barnes LD (1974) Effects of colchicine and vinblastine on the cellular action of vasopressin in the mammalian kidney. A possible role of microtubles. J Clin Invest 54:252–262

Dousa TP (1999) Cyclic-3',5'-nucleotide phosphodiesterase isozymes in the cell biology and pathophysiology of the kidney. Kidney Int 55:29–62

Dransfield DT, Bradford A J, Smith J, Martin M, Roy C, Mangeat PH, Goldenring JR (1997) Ezrin is a cyclic AMP-dependent protein kinase anchoring protein. EMBO J 16:35–43

Earm JH, Christensen BM, Frokiaer J, Marples D, Han JS, Knepper MA, Nielsen S (1998) Decreased aquaporin-2 expression and apical plasma membrane delivery in kidney collecting ducts of polyuric hypercalcemic rats. J Am Soc Nephrol 9:2181–2193

Ecelbarger CA, Chou C-L, Lolait SJ, Knepper MA, DiGiovanni SR (1996) Evidence for dual signaling pathways for V_2 vasopressin receptor in rat inner medullary collecting duct. Am J Physiol 270:F623–F633

Ecelbarger CA, Nielsen S, Olson BR, Murase T, Baker EA, Knepper MA, Verbalis JG (1997) Role of renal aquaporins in escape from vasopressin-induced antidiuresis in rat. J Clin Invest 99:1852–1863

Ecelbarger CA, Chou CL, Lee AJ, DiGiovanni S, Verbalis JG, Knepper MA (1998) Escape from vasopressin-induced antidiuresis:role of vasopressin resistance of the collecting duct. Am J Physiol 274:F1161–F1166

Echard A, Jollivet F, Martinez O, Lacapere JJ, Rousselet A, Janoueix-Lerosey I, Goud B (1998) Interaction of a Golgi-associated kinesin-like protein with Rab6. Science 279:580–585

Edwards RM, Stack EJ, Pullen M, Nambi P (1993) Endothelin inhibits vasopressin action in rat inner medullary collecting duct via the ET_B receptor. J Pharmacol Exp Ther 267:1028–1033

Edwards RM, Spielman WS (1994) Adenosine A1 receptor mediated inhibition of vasopressin action in inner medullary collecting duct. Am J Physiol 266:F791–F796

Edwardson JM, An S, Jahn R (1997) The secretory granule protein syncollin binds to syntaxin in a Ca2(+)-sensitive manner. Cell 90:325–333

Eker P, Holms PK, van Deurs B, Sandvig K (1994) Selective regulation of apical endocytosis in polarized madin-darby canine kidney cells by mastoparan and cAMP. J Biol Chem 269:18607–18615

Emma F, Harris HW, Strange K (1994) Acidification of vasopressin-induced endosomes in toad urinary bladder. Am J Physiol 267:F106–F113

Emmons C (1999) Transport characteristics of the apical anion exchanger of rabbit cortical collecting duct β-cells. Am J Physiol 276:F635–F643

Fassauer D, Antonin W, Margittai M, Pabst S, Jahn R (1999) Mixed and non-cognate SNARE complexes. J Biol Chem 274:15440–15446

Faux MC, Scott JD (1996) Molecular glue:kinase anchoring and scaffold proteins. Cell 85:9–12

Ferro-Novick S, Jahn R (1994) Vesicle fusion from yeast to man. Nature 370:191–193.

Fischer von Mollard G, Südhof TC, Jahn R (1991) A small GTP-binding protein dissociates from synaptic vesicles during exocytosis. Nature 349:79–81

Fischer von Mollard G, Stahl B, Khokhlatchev A, Sudhof TC, Jahn R (1994a) Rab3C is a synaptic vesicle protein that dissociates from synaptic vesicles after stimulation of exocytosis. J Biol Chem 269:10971–10974

Fischer von Mollard G, Stahl B, Walch-Solimena C, Takei K, Daniels L, Khoklatchev A, De Camilli P, Südhof TC, Jahn R (1994b) Localization of Rab5 to synaptic vesicles identifies endosomal intermediate in synaptic vesicle recycling pathway. Eur J Cell Biol 65:319–326

Franki N, Macaluso F, Gao Y, Hays RM (1995a) Vesicle fusion proteins in rat inner medullary collecting duct and amphibian bladder. Am J Physiol 268:C792–797

Franki N, Macaluso F, Schubert W, Gunther L, Hays RM (1995b) Water channel-carrying vesicles in the rat IMCD contain cellubrevin. Am J Physiol 269:C797–801

Friis UG, Jensen BL, Aas JK, Skott O (1999) Direct demonstration of exocytosis and endocytosis in single mouse juxtaglomerular cells. Circ Res 84:929–936

Frokiaer J, Marples D, Knepper MA, Nielsen S (1996) Bilateral ureteral obstruction downregulates expression of vasopressin-sensitive AQP2 water channel in rat kidney. Am J Physiol 270:F657–668

Frokiaer J, Christensen CM, Marples D, Djurhuus JC, Jensen UB, Knepper MA, Nielsen S (1997) Downregulation of aquaporin-2 parallels changes in renal water excretion in unilateral ureteral obstruction. Am J Physiol 273:F213–F223

Fujita N, Ishikawa SE, Saski S, Fujisawa G, Fushimi K, Marumo F, Saito S (1995) Role of water channel AQP-CD in water retention in SIADH and cirrhotic rats. Am J Physiol 269:F926–F931

Fujita-Yoshigaki J, Dohke Y, Hara-Yokoyama M, Kamata Y, Kozaki S, Furuyama S, Sugiya H (1996) Vesicle-associated membrane protein 2 is essential for cAMP-regulated exocytosis in rat parotid acinar cells:the inhibition of cAMP-dependent amylase release by botulinum neurotoxin B. J Biol Chem 271:13130–13134

Fujita-Yoshigaki J (1998) Divergence and convergence in regulated exocytosis:the characteristics of cAMP-dependent enzyme secretion of parotid salivary acinar cells. Cell Signal 10:371–375

Furuno M, Uchida S, Marumo F, Sasaki S (1996) Repressive regulation of the AQP2 gene. Am J Physiol 271:F854–F860

Fushimi K, Uchida S, Hara Y, Hirata Y, Marumo F, Sasaki S (1993) Cloning and expression of apical membrane water channel of rat kidney collecting tubule. Nature 361:549–552

Fushimi K, Sasaki S, Marumo F (1997) Phosphorylation of serine 256 is required for cAMP-dependent regulatory exocytosis of the aquaporin-2 water channel. J Biol Chem 272:14800–14804

Garg LC, Kapturczak E (1990) Stimulation of phosphoinositide hydrolysis in renal medulla by vasopressin. Endocrinology 127:1022–1027

Garrett JE, Capuano IV, Hammerland LG, Hung BC, Brown EM, Hebert SC, Nemeth EF, Fuller F (1995) Molecular cloning and functional expression of human parathyroid calcium receptor cDNAs. J Biol Chem 270:12919–12925

Gasman S, Chasserot-Golaz S, Hubert P, Aunis D, Bader MF (1998) Identification of a potential effector pathway for the trimeric G_o protein associated with secretory granules. G_o stimulates a granule-bound phosphatidyl 4-kinase by activating RhoA in chromaffin cells. J Biol Chem 273:16913–16920

Geppert M, Bolshakov VY, Siegelbaum SA, Takei E, DeCamilli P, Hammer RE, Südhof TC (1994) The role of rab3A in neurotransmitter release. Nature 369:493–497

Geppert M, Goda Y, Stevens CF, Südhof TC (1997) The small GTP-binding protein Rab3A regulates a late step in synaptic vesicle fusion. Nature 387:810–814

Geppert M, Südhof TC (1998) Rab3 and synaptotagmin:the yin and yang of synaptic transmission. Annu Rev Neurosci 21:75–95

Gill SR, Schroer TA, Szilak I, Steuer ER, Sheetz MP, Cleveland DW (1991) Dynactin, a conserved, ubiquitously expressed component of an activator of vesicle motility mediated by cytoplasmic dynein. J Cell Biol 115:1639–1650

Goldenring JR, Shen KR, Vaughan HD, Modlin IM (1993) Identification of a small GTP-binding protein, Rab25, expressed in the gastrointestinal mucosa, kidney, and lung. J Biol Chem 268:18419–18422

Goldstein JL, Anderson RGW, Brown MS (1979) Coated pits, coated vesicles, and receptor mediated endocytosis. Nature 279:679–685

Gonzalez Jr L, Scheller RH (1999) Regulation of membrane trafficking:structural insights from a rab/effector complex. Cell 96:755–758

Grantham JJ, Burg MB (1966) Effect of vasopressin and cyclic AMP on permeability of isolated collecting tubules. Am J Physiol 211:255-259

Gruenberg J, Maxfield FR (1995) Membrane transport in the endocytic pathway. Curr Opin Cell Biol 7:5563

Guan Y, Zhang Y, Breyer RM, Fowler B, Davis L, Hebert RL, Breyer MD (1998) Prostaglandin E2 inhibits renal collecting duct Na^+ absorption by activating the EP1 receptor. J Clin Invest 1998 102:1–201

Gustafson CE, Levine S, Katsura T, McLaughlin M, Aleixo MD, Tamarappoo BK, Verkman AS, Brown D (1998) Vasopressin-regulated trafficking of a green fluorescent protein-aquaporin2 chimera in LLC-PK$_1$ cells. Histochem Cell Biol 110:377–386

Hackenthal E, Paul M, Ganten D, Taugner R (1990) Morphology, physiology, and molecular biology of renin secretion. Physiol Rev 70:1076–1116

Hadley ME (1996) Endocrinology. Prentice Hall International, Inc. NJ, USA

Hall A (1998) Rho GTPases and the actin cytoskeleton. Science 279:509–514

Hamm-Alvarez SF, Sheetz MP (1998) Microtubule-dependent vesicle transport:modulation of channel and transporter activity in liver and kidney. Phys Rev 78:1109–1129

Hammond TG, Morre DJ, Harris Jr HW, Zeidel ML (1993) Isolation of highly purified, functional endosomes from toad urinary bladder. Biochem J 295:471–476

Handler JS (1988) Antidiuretic hormone moves membranes. Am J Physiol 255:F375–F382

Harmanci MC, Stern P, Kachadorian WA, Valtin H, DiScala VA (1978) Antidiuretic hormone-induced intramembraneous alterations in mammlian collecting duct. Am J Physiol 235 F440–F443

Harris Jr HW, Strange K, Zeidel ML (1991) Current understanding of the cellular biology and molecular structure of the antidiuretic hormone-stimulated water transport pathway. J Clin Invest 88:1–8

Harris HW Jr, Zeidel ML, Jo I, Hammond TG (1994) Characterization of purified endosomes containing the antidiuretic hormone-sensitive water channel from rat renal papilla. J Biol Chem 269:11993–2000

Hasegawa H, Ma T, Skach W, Matthay MA, Verkman AS (1994) Molecular cloning of a mercurial insensitive water channel expressed in selected water-transporting tissues. J Biol Chem 269:5497–5500

Hay JC, Scheller RH (1997) SNAREs and NSF in targeted membrane fusion. Curr Opin Cell Biol 9:505–512

Hays RM, Leaf A (1962) Studies on the movement of water through the isolated toad bladder and its modification by vasopressin. J Gen Physiol 45:905–919

Hays RM, Franki N, Simon H, Gao Y (1994) Antidiuretic hormone and exocytosis:lessons from neurosecretion. Am J Physiol 267:C1507–C1524

Hays, R.M. (1996) Cellular and molecular events in the action of antidiuretic hormone. Kindney Int 49, 1700–1705

Hebert RL, Jacobson HR, Breyer MD (1990) PGE$_2$ inhibits AVP-induced water flow in cortical collecting ducts by protein kinase C activation. Am J Physiol 259:F318–325.

Hebert RL, Jacobson HR, Breyer MD (1991) Prostaglandin E2 inhibits sodium transport in rabbit cortical collecting duct by increasing intracellular calcium. J Clin Invest 87:1992–1998

Hebert RL, Jacobson HR, Fredin D, Breyer MD (1993) Evidence that separate PGE$_2$ receptors modulate water and sodium transport in rabbit cortical collecting duct. Am J Physiol 265:F643–650

Helms JB (1995) Role of heterotrimeric GTP binding proteins in vesicular protein transport:indications for both classical and alternative G protein cycles. FEBS Lett 369:84–88

Helms JB, Helms-Brons D, Brugger B, Gkantiragas I, Eberle H, Nickel W, Nurnberg B, Gerdes HH, Wieland FT (1998) A putative heterotrimeric G protein inhibits the fusion of COPI-coated vesicles. Segregation of heterotrimeric G proteins from COPI-coated vesicles. J Biol Chem 273:15203–15208

Holleran EA, Karki S, Holzbaur ELF (1998) The role of dynactin complex in intracellular motility. Int Rev Cytol 182:69–109

Höhler T, Teuber G, Warnitschke R, Meyer zum Büschenfeld KH (1994) Indomethacin treatment in amphotericin B induced neprogenic diabetes insipidus. Clin Invest 72:769–771

Huang LJ, Durick K, Weiner JA, Chun J, Taylor SS (1997a) Identification of a novel protein kinase A anchoring protein that binds both type I and type II regulatory subunits. J Biol Chem 272:8057–8064

Huang LJ, Durick K, Weiner JA, Chun J, Taylor SS (1997b) D-AKAP2, a novel protein kinase A anchoring protein with a putative RGS domain. Proc Natl Acad Sci USA 94:11184–1118

Inoue T, Nielsen S, Mandon B, Terris J, Kishore BK, Knepper M (1998) SNAP-23 in rat kidney:colocalization with aquaporin-2 in collecting duct vesicles. Am J Physiol 275:F752–F760

Ishibashi K, Sasaki S, Fushimi K, Uchida S, Kuwahara M, Saito H, Furukawa T, Nakajima K, Yamaguchi Y, Gojobori T, Marumo F (1994) Molecular cloning and expression of a member of the aquaporin family with peremeability to glycerol and urea in addition to water expressed at the basolateral membrane of kidney collecting duct cells. Proc Natl Acad Sci USA 91:6269–6273

Jahn R, Südhof TC (1994) Synaptic vesicles and exocytosis. Annu Rev Neurosci 17:219–46

Jamora C, Takizawa PA, Zaarour RF, Denesvre C, Faulkner J, Malhotra V (1997) Regulation of Golgi structure through heterotrimeric G proteins. Cell 91:617–626

Jena BP, Schneider SW, Geibel JP, Webster P, Oberleithner H, Sritharan KC (1997) Gi regulation of secretory vesicle swelling examined by atomic force microscopy. Proc Natl Acad Sci USA 94:13317–13322

Jo I, Harris HW, Amendt-Raduege AM, Majewski RR, Hammond TG (1995) Rat kidney papilla contains abundant synaptobrevin protein that participates in the fusion of antidiuretic hormone-regulated water channel-containing endosomes in vitro. Proc Natl Acad Sci USA 92:1876–1880

Jung JS, Bhat RV, Preston GM, Guggino WB, Baraban JM, Agre P (1994) Molecular characterization of an aquaporin cDNA from brain:candidate osmoreceptor and regulator of water balance. Proc Natl Acad Sci USA 91:13052–13056

Just I, Selzer J, Wilm M, von Eichel-Streiber C, Mann M, Aktories K (1995) Glucosylation of Rho proteins by Clostridium difficile toxin B. Nature 375:500–503

Kachadorian WA, Sariban-Sohraby S, Spring KR (1985) Regulation of water permeability in toad urinary bladder at two barriers. Am J Physiol 248:F260–F265

Kamsteeg EJ, Wormhoudt TA, Rijss JP, van Os CH, Deen PM (1999). An impaired routing of wild-type aquaporin-2 after tetramerization with an aquaporin-2 mutant explains dominant nephrogenic diabetes insipidus. EMBO J 18:2394–2400

Karki S, Holzbaur ELF (1999) Cytoplasmic dynein and dynactin in cell division and intracellular transport. Curr Opin Cell Biol 11:45–53

Kasner M, Grosse J, Krebs M, Kaczmarczyk G (1995) Methohexital impairs osmoregulation:studies in concsious and anesthetized volume-expanded dogs. Anesthesiology 82:1396–1405

Kato M, Sasaki T, Ohya T, Nakanishi H, Nishioka H, Imamura M, Takai Y (1996) Physical and functional interaction of rabphilin-3A with alpha-actinin. J Biol Chem 271:31775–31778

Katsura T, Verbavatz JM, Farinas J, Ma T, Ausiello DA, Verkman AS, Brown D (1995) Constitutive and regulated membrane expression of aquaporin1 and aquaporin2 water channels in stably transfected LLC-PK$_1$ epithelial cells. Proc Natl Acad Sci USA 92:7212–7216

Katsura T, Ausiello DA, Brown D (1996) Direct demonstration of aquaporin-2 water channel recycling in stably transfected LLC-PK$_1$ epithelial cells. Am J Physiol 270:F548–F553

Katsura T, Gustafson CE, Ausiello DA, Brown D (1997) Protein kinase A phosphorylation is involved in regulated exocytosis of aquaporin-2 in transfected LLC-PK$_1$ cells. Am J Physiol 272:F816–F822

Kehlenbach RH, Matthey J, Huttner WB (1994) Xl alpha s is a new type of G protein. Nature 372:804–809. (1995) 375:253

Klussmann E, Maric K, Krause E, Eichhorst J, Beziat P, Rosenthal W (1998) Molecular mechanisms of exocytosis in epithelial cells of the kidney. Congress of Molecular Medicine. Berlin, Germany, 1998. J Mol Med 76:B 58 (abstract)

Klussmann E, Maric K, Wiesner B, Beyermann M, Rosenthal W (1999) Protein kinase A anchoring proteins are required for vasopressin-mediated translocation of aquaporin-2 into cell membranes of renal principal cells. J Biol Chem 274:4934–4938

Knepper MA, Wade JB, Terris J, Ecelbarger CA, Marples D, Mandon B, Chou CL, Kishore BK, Nielsen S (1996) Renal aquaporins. Kidney Int 49:1712–1717

Koeford-Johnson V, Ussing HH (1953) The contributions of diffusion and flow to the passage of D$_2$O through living membranes. Acta Physiol Scand 28:60–76

Kohan DE, Padilla E, Hughes AK (1993) Endothelin B receptor mediates ET-1 effects on cAMP and PGE$_2$ accumulation in rat IMCD. Am J Physiol 265:F670–F676

Kohan DE (1998) Endothelin modulation of renal sodium and water transport. In: Pollock DM, Highsmith RF (eds) Endothelin receptors and signaling mechanisms. Springer-Verlag and R.G. Landes Company, pp 101–113

Konrad RJ, Young RA, Record RD, Smith RM, Butkerait P, Manning D, Jarret L, Wolf BA (1995) The heterotrimeric G protein G$_i$ is localized to the insulin secretory granules of beta-cells and is involved in insulin exocytosis. J Biol Chem 270:12869–12876

Kowluru A, Seavey SE, Rhodes CJ, Metz SA (1996) A novel regulatory mechanism for trimeric GTP-binding proteins in the membrane and secretory granule fractions of human and rodent beta cells. Biochem J 313:97–107

Kumagami H, Loewenheim H, Beitz E, Wild K, Schwartz H, Yamashita K, Schultz J, Paysan J, Zenner HP, Ruppersberg JP (1998) The effect of antidiuretic hormone on the endolymphatic sac of the inner ear. Pflugers Arch 436:970–975

Kurtz A (1989) Cellular control of renin secretion. Rev Physiol Biochem Pharmacol 113:1–40

Kuwahara M, Fushimi K, Terada Y, Bai L, Marumo F, Sasaki S (1995) cAMP-dependent phosphorylation stimulates water permeability of aquaporin-collecting duct water channel protein expressed in Xenopus oocytes. J Biol Chem 270:10384–10387

Kwon TH, Frokiaer J, Knepper MA, Nielsen S (1998) Reduced AQP1, -2, and -3 levels in kidneys of rats with CRF induced by surgical reduction in renal mass. Am J Physiol 275:F724–F741

Lambright DG, Sondek J, Bohm A, Skiba NP, Hamm HE, Sigler PB (1996) The 2.0 A crystal structure of a heterotrimeric G protein. Nature 379:311–319

Lande MB, Jo I, Zeidel ML, Somers M, Harris HW (1996) Phosphorylation of aquaporin-2 does not alter the membrane water permeability of papillary water channel-containing vesicles. J Biol Chem 271:5552–5557

Lester LB, Langeberg LK, Scott JD (1997) Anchoring of protein kinase A facilitates hormone-mediated insulin secretion. Proc Natl Acad Sci USA 94:14942–14947

Lester LB, Scott JD (1997) Anchoring and scaffold proteins for kinases and phosphatases. Recent Prog Horm Res 52:409–430

Li C, Takei K, Geppert M, Daniell L, Stenius K, Chapman ER, Jahn R, DeCamilli P, Südhof TC (1994) Synaptic targeting of rabphilin-3A, a synaptic vesicle Ca^{2+}/phospholipid-binding protein, depends on rab3A/3C. Neuron 13:885–98

Liebenhoff U, Rosenthal W (1995) Identification of Rab3, Rab5a- and synaptobrevin II-like proteins in a preparation of rat kidney vesicles containing the vasopressin-regulated water channel. FEBS Lett 365:209–213

Linial M (1997) SNARE proteins - why so many, why so few? J Neurochem 69:1781–1792

Lledo PM, Vernier P, Vincent JD, Mason WT, Zorec R (1993) Inhibition of Rab3B expression attenuates Ca(2+)-dependent exocytosis in rat anterior pituitary cells. Nature 364:540–544

Ludger J, Galli T (1998) Exocytosis:SNAREs drum up. Eur J Neurosci 10:415–422

Lutcke A, Jansson S, Parton RG, Chavrier P, Valencia A, Huber LA, Lehtonen E, Zerial M (1993) Rab17, a novel small GTPase, is specific for epithelial cells and is induced during cell polarization. J Cell Biol 121:553–564

Luzius H, Jans DA, Grünbaum EG, Moritz A, Rascher W, Fahrenholz F (1992) A low affinity vasopressin V2-receptor in inherited nephrogenic diabetes insipidus. J Recept Res 12:351–368

Ma T, Frigeri A, Skach W, Verkman AS (1993a) Cloning of a novel rat kidney cDNA homologous to CHIP28 and WCH-CD water channels. Biochem Biophys Res Comm 197:654–659

Ma T, Frigeri A, Tsai ST, Verbavatz JM, Verkman AS (1993b) Localization and functional analysis of CHIP28k water channels in stably transfected chinese hamster ovary cells. J Biol Chem 268:22756–22764

Ma T, Yang B, Gillespie A, Carlson EJ, Verkman AS (1998) Severly impaired urinary concentrating ability in transgenic mice lacking aquaporin-1 water channels. J Biol Chem 273:4296–4299

Maeda Y, Han JS, Gibson CC, Knepper MA (1993) Vasopressin and oxytocin receptors coupled to Ca^{2+} mobilization in rat inner medullary collecting duct. Am J Physiol 265:F15–F25

Mandelkow E, Hoenger A (1999) Structures of kinesin and kinesin-microtubule interactions. Curr Opin Cell Biol 11:34–44

Mandon B, Chou C-L, Nielsen S, Knepper M (1996) Syntaxin-4 is localized to the apical plasma membrane of rat renal collecting duct cells:possible role in aquaporin-2 trafficking. J Clin Invest 98:906–913

Mangeat P, Roy C, Martin M (1999) ERM proteins in cell adhesion and membrane dynamics. Trends Cell Biol 9:187–192

Maric K, Oksche A, Rosenthal W (1998) Aquaporin-2 expression in primary cultured rat inner medullary collecting duct cells. Am J Physiol 275:F796–801

Marples D, Christensen S, Christensen EI, Ottosen PD, Nielsen S (1995) Lithium-induced downregulation of aquaporin-2 water channel expression in rat kidney medulla. J Clin Invest 95:1838–1845

Marples D, Frokiaer J, Dorup J, Knepper MA, Nielsen S (1996) Hypokalemia-induced downregulation of aquaporin-2 water channel expression in rat kidney medulla and cortex. J Clin Invest 97:1960–1968

Marples D, Schroer TA, Ahrens N, Taylor A, Knepper MA, Nielsen S (1998) Dynein and dynactin colocalize with AQP2 water channels in intracellular vesicles from kidney collecting duct. Am J Physiol 274:F384–F394

Marples D, Froekiaer J, Nielsen S (1999) Long-term regulation of aquaporins in the kidney. Am J Physiol 276:F331–339

Martinez O, Goud B (1998) Rab proteins. Biochim Biophys Acta 1404:101–112

McMahon HT, Ushkaryov YA, Edelmann L, Link E, Binz T, Niemann H, Jahn R, Südhoff TC (1993) Cellubrevin is a ubiquitous tetanus-toxin substrate homologous to a putative synaptic vesicle fusion protein. Nature 364:346–349

McLauchlan H, Newell J, Morrice N, Osborne A, West M, Smythe E (1998) A novel role for Rab5-GDI in ligand sequestration into clathrin-coated pits. Curr Biol 8:34–45

Meads T, Schroer TA (1995) Polarity and nucleation of microtubules in polarized epithelial cells. Cell Motil Cytoskeleton 32:273–288

Migas I, Backer A, Meyer-Lehnert H, Michel H, Wulfhekel U, Kramer HJ (1993) Characteristics of endothelin receptors and intracellular signalling in porcine inner medullary collecting duct cells. Am J Hypertens 6:611–618

Mills IG, Jones AT, Clague MJ (1999) Regulation of endosome fusion. Mol Membr Biol 16:73–79

Montmayeur J-P, Borrelli E (1994) Targeting of $G_{\alpha i2}$ to the Golgi by alternative spliced carboxyl-terminal region. Science 263:95–98

Mohrmann K, van der Sluijs P (1999) Regulation of membrane transport through the endocytic pathway by rabGTPases. Mol Membr Biol 16:81–87

Mork A, Geisler A (1995) A comparative study on the effects of tetracyclines and lithium on the cyclic-AMP and second messenger system in the rat brain. Prog Neuro Psycho Biol Psych 19:157–169

Moullier P, Friedlander G, Calise D, Ronco P, Perricaudet M, Ferry N (1994) Adenoviral-mediated gene transfer to renal tubular cells *in vivo*. Kidney Int 45:1220–1225

Mulders SM, Bichet DG, Rijss JP, Kamsteeg EJ, Arthus MF, Lonergan M, Fujiwara M, Morgan K, Leijendekker R, van der Sluijs P, van Os CH, Deen PM (1998) An aquaporin-2 water channel mutant which causes autosomal dominant nephrogenic diabetes insipidus is retained in the Golgi complex. J Clin Invest 102:57–66

Niederberger M, Martin PY, Gines P (1995) Normalization of nitric oxide production corrects arterial vasodilatation and hyperdynamic circulation in cirrhotic rats. Gastroenterology 109:1624–1630

Nielsen S, Muller J, Knepper MA (1993a) Vasopressin- and cAMP-induced changes in ultrastructure of isolated perfused inner medullary collecting ducts. Am J Physiol 265:F225–238

Nielsen S, DiGiovanni SR, Christensen EI, Knepper MA, Harris HW (1993b) Cellular and subcellular immunolocalization of vasopressin-regulated water channel in rat kidney. Proc Natl Acad Sci USA 90:11663–11667

Nielsen S, Marples D, Birn H, Mohtashami M, Dalby NO, Trimble W, Knepper MA (1995a) Expression of Vamp2-like protein in kidney collecting duct intracellular vesicles. J Clin Invest 96:1834–1844

Nielsen S, Chou C-L, Marples D, Christensen EI, Kishore BK, Knepper MA (1995b) Vasopressin increases water permeability of kidney collecting duct by inducing translocation of aquaporin-CD water channels to plasma membrane. Proc Natl Acad Sci USA 92:1013–1017

Nielsen S, Terris J, Andersen D, Ecelbarger C, Frokiaer J, Jonassen T, Marples D, Knepper MA, Petersen JS (1997) Congestive heart failiure in rats associated with increased expression and targetting of aquaporin-2 water channel in collecting duct. Proc Natl Acad Sci USA 94:5450–5455

Nonoguchi H, Sands JM, Knepper MA (1988) Atrial natriuretic factor inhibits vasopressin-stimulated osmotic water permeability in rat inner medullary collecting duct. J Clin Invest 82:1383–1390

Nonoguchi H, Sands JM, Knepper MA (1989) ANF inhibits NaCl and fluid absorption in cortical collecting duct of rat kidney. Am J Physiol 256:F179–F186

Novick P, Zerial M (1997) The diversity of Rab proteins in vesicle transport. Curr Opin Cell Biol 9:496–504

Nürnberg B, Ahnert-Hilger G (1996) Potential roles of heterotrimeric G proteins of the endomembrane system. FEBS Lett 389:61–65

Offermann S, Simon M (1996) Organization of transmembrane signalling by heterotrimeric G proteins. In:Cancer Surveys 27:Cell Signalling. Imperial Cancer Research Fund

Ohara M, Martin PY, Xu DL, StJohn J, Pattison TA, Kim JK, Schrier RW (1998) Upregulation of aquaporin 2 water channel expression in pregnent rats. J Clin Invest 101:1076–1083

Oksche A, Rosenthal W (1998) The molecular basis of nephrogenic *diabetes in-sipidus.* J Mol Med 76:326–337

Olefsky JM (1999) Insulin-stimulated glucose transport minireview series. J Biol Chem 274:1863

Osborne SL, Herreros J, Bastiaens PIH, Schiavo G (1999) Calcium-dependent oligomerization of synaptotagmins I and II. J Biol Chem 274:59–66

Otto H, Hanson PI, Jahn R (1997) Assembly and disassembly of a ternary complex of synaptobrevin, syntaxin, and SNAP-25 in the membrane of synaptic vesicles. Proc Natl Acad Sci USA 94:6197–6201

Palmer LG, Lorenzen M (1983) Antidiuretic hormone-dependent membrane capacitance and water permeability in the toad urinary bladder. Am J Physiol 244:F195–F204

Pawson T, Scott JD (1997) Signalling through scaffolding, anchoring and adaptor proteins. Science 278:2075–2080

Pearl M, Taylor A (1983) Actin filaments and vasopressin-stimulated water flow in toad urinary bladder. Am J Physiol 245:C28–C39

Peters C, Mayer A (1998) Ca^{2+}/calmodulin signals the completions of docking and triggers a late step of vacuole fusion. Nature 396:575–580

Pfeffer SR (1999) Transport-vesicle targeting:tethers before SNAREs. Nature Cell Biol 1:E17–E22

Phillips ME, Taylor A (1989) Effect of nocodazole on the water permeability response to vasopressin in rabbit collecting tubules in vitro. J Physiol 411:529–544

Phillips ME, Taylor A (1992) Effect of colcemid on the water permeability response to vasopressin in isolated perfused rabbit collecting tubules. J Physiol 456:591–608

Pimplikar SW, Simons K (1994) Activators of protein kinase A stimulate apical but not basolateral transport in epithelial madin-darby canine kidney cells. J Biol Chem 269:19054–19059

Pinxteren JA, O'Sullivan AJ, Tatham PER, Gomberts BD (1998) Regulation of exocytosis from rat peritoneal mast cells by G protein βγ-subunits. EMBO J 17:6210–6218

Preston GM, Agre P. (1991) Isolation of the cDNA for erythrocyte integral membrane protein of 28 kilodaltons:Member of an ancient channel family. Proc Natl Acad Sci USA 88:11110–11114

Preston GM, Carroll TP, Guggino WB, Agre P (1992) Appearence of water channels in *Xenopus* oocytes expressing red cell CHIP28 protein. Science 256:385–387

Preston GM, Smith BL, Christensen EI, Agre P (1994) Mutations in aquaporin-1 in phenotypically normal humans without functional CHIP water channels. Science 265:1585–1587

Rai T, Uchida S, Marumo F, Saski S (1997) Cloning of rat and mouse aquaporin-2 gene promoters and identification of a negative *cis*-regulatory element. Am J Physiol 273:F264–F273

Ravichandra V, Chawla A, Roche PA (1996) Identification of a novel syntaxin- and synaptobrevin/Vamp-binding protein, SNAP-23, expressed in non-neuronal cells. J Biol Chem 271:13300–13303

Rea S, Martin LB, McIntosh S, Macaulay SL, Ramsdale T, Baldini G, James DE (1998) Syndet, an adipocyte target SNARE involved in the insulin-induced translocation of GLUT4 to the cell surface. J Biol Chem 273:18784–18792

Ren M, Xu G, Zeng J, de Lemos-Chiarandini C, Adesnik M, Sabatini DD (1998) Hydrolysis of GTP on rab11 is required for direct delivery of transferrin from the pericentriolar recycling compartment to the cell surface but not from sorting endosomes. Proc Natl Acad Sci USA 95:6187–6192

Riezman H, Woodman PG, van Meer G, Marsh M (1997) Molecular mechanism of endocytosis. Cell 91:731–738

Robertson GL (1995) Diabetes insipidus. Endocrinol Metab Clin North Am. 24:549–72

Robinson LJ, Martin TFJ (1998) Docking and fusion in neurosecretion. Curr Opin Cell Biol 10:483–492

Rosenthal W, Seibold A, Antaramian A, Lonergan M, Arthus M-F, Hendy G, Birnbaumer M, Bichet D (1992) Molecular identification of the gene responsible for congenital nephrogenic diabetes insipidus. Nature 359:233–235

Rosenthal W, Seibold A, Bichet DG, Birnbaumer M (1996) Diabetes insipidus. In: Encyclopedia of Molecular Biology and Molecular Medicine 2:19–28

Rosenthal W, Oksche A, Bichet DG (1998) Two genes-one disease:the molecular basis of congenital nephrogenic diabetes insipidus. Adv Mol Cell Endocrinol 2:143–167

Rothman JE (1994) Mechanisms of intracellular protein transport. Nature 372:55-63

Roychowdhury S, Rasenick MM (1997) G protein beta1gamma2 subunits promote microtubule assembly. J Biol Chem 272:31576–31581

Roychowdhury S, Panda D, Wilson L, Rasenick MM (1999) G protein α subunits activate tubulin GTPase and modulate microtubule polymerization dynamics. J Biol Chem 274:13485–13490

Rubin CS (1994) A kinase anchor proteins and the intracellular targeting of signals carried by cyclic AMP. Biochim Biophys Acta 1224:467–479

Ruhfus B, Bauernschmitt HG, Kinne RKH (1998) Properties of a polarized primary culture from rat renal inner medullary collecting duct (IMCD) cells. In Vitro Cell Dev Biol – Animal 34:227–231

Rutishauser J, Kopp P (1999) Aquaporin-2 water channel mutations and nephrogenic diabetes insipidus:new variations on a theme. Eur J Endocrinol 140:137–139

Rybin V, Ullrich O, Rubino M, Alexandrov K, Simon I, Seabra MC, Goody R, Zerial M (1996) GTPase activity of Rab5 acts as a timer for endocytic membrane fusion. Nature 383:266–269

Saavedra JM (1992) Brain and pituitary angiotensin. Endocr Rev 13:329–380

Sabolic I, Wuarin F, Shi LB, Verkman AS, Ausiello DA, Gluck S, Brown D (1992) Apical endosomes isolated from kidney collecting duct principal cells lack subunits of the proton pumping ATPase. J Cell Biol 119:111–22

Sabolic I, Katsura T, Verbavatz J-M, Brown D (1995) The AQP2 water channel: effect of vasopressin treatment, microtubule disruption, and distribution in neonatal rats. J Mem Biol 143:165–175

Sakairi Y, Jacobson HR, Noland TD, Breyer MD (1995) Luminal prostaglandin E receptors regulate salt and water transport in rabbit cortical collecting duct. Am J Physiol 269:F257–F265

Sands JM, Naruse M, Baum M, Jo I, Hebert SC, Brown EM, Harris HW (1997) Apical extracellular calcium/polyvalent cation-sensing receptor regulates vasopressin-elicited water permeability in rat kidney inner medullary collecting duct. J Clin Invest 99:1399–1405

Sasaki S, Fushimi K, Saito H, Saito F, Uchida S, Ishibashi K, Kuwahara M, Ikeuchi T, Inui K, Nakajama K,Watanabe TX, Marumo F. (1994) Cloning, characterization, and chromosomal mapping of human aquaporin of collecting duct. J Clin Invest 93:1250–1256

Sasaki S, Ishibashi K, Marumo F (1998) Aquaporin-2 and -3:representatives of two subgroups of the aquaporin family colocalized in the kidney collecting duct. Annu Rev Physiol 60:199–220

Schimmöller F, Simon I, Pfeffer SR (1998) Rab GTPases, directors of vesicle docking. J Biol Chem 273:22161–22164

Schliefer K, Rockstroh JK, Spengler U, Sauerbruch T (1997) Nephrogenic diabetes insipidus in a patient taking cidofovir. Lancet 350:413–414; 350:1558

Schmale H, Richter D (1984) Single base deletion is the cause of diabetes insipidus in Brattleboro rats. Nature 308:705–709

Schmale H, Bahnsen U, Richter D (1993) Structure and expression of the vasopressin precursor gene in central diabetes insipidus. Ann NY Acad Sci 689:74–82

Schmid SL (1997) Clathrin-coated vesicle formation and protein sorting:an integrated process. Annu Rev Biochem 66:511–548

Schmidt G, Aktories K (1998) Bacterial cytotoxins target Rho GTPases. Naturwissenschaften 85:253–261

Schrier RW, Fassett RG, Ohara M, Martin PY (1998) Vasopressin release, water channels, and vasopressin antagonism in cardiac failiure, cirrhosis and pregnancy. Proc Assoc Am Physicians 110:407–411

Schroer TA, Sheetz MP (1991) Two activators of microtubule-based vesicle tranport. J Cell Biol 115:1309–1318

Schülein R, Lorenz D, Oksche A, Wiesner B, Hermosilla R, Ebert J, Rosenthal W (1998) Polarized cell surface expression of the green fluorescent-protein vasopressin V2 receptor in Madin Darby canine kidney cells. FEBS Lett 441:170–176

Scott JD (1997) Dissection of protein kinase and phosphatase targeting interactions. Soc Gen Physiol Ser 52:227–239

Seasholtz TM, Majumdar M, Brown JH (1999) Rho as mediator of G protein-coupled receptor signalling. Mol Pharmacol 55:949–956

Sever S, Mulberg AB, Schmid SL (1999) Impairment of dynamin's GAP domain stimulates receptor-mediated endocytosis. Nature 398:481–486

Shirataki H, Kaibuchi K, Sakoda T, Kishida S, Yamaguchi T, Wada K, Miyazaki M, Takai Y (1993) Rabphilin-3A, a putative target protein for smg p25A/rab3A p 25 small GTP-binding protein related to synaptotagmin. Mol Cell Biol 13:2061–2068

Simon H, Gao Y, Franki N, Hays RM (1993) Vasopressin depolymerizes apical F-actin in rat inner medullary collecting duct. Am J Physiol 265:C757–C762

Slembrouck D, Annaert WG, Wang JM, Potter WP (1999) Rab3 is present on endosomes from bovine chromaffin cells in primary culture. J Cell Sci 112:641–649

Snyder HM, Fredin DM, Breyer MD (1991) Muscarinergic receptor activation inhibits AVP-induced water flow in rabbit cortical collecting ducts. Am J Physiol 260:F929–F936

Snyder HM, Noland TD, Breyer MD (1992) cAMP-dependent protein kinase mediates hydrosmotic effect of vasopressin in collecting duct. Am J Physiol 263:C147–C153

Sonnenberg H (1990) Mechanisms of release and renal action of atrial natriuretic factor. Acta Physiol Scand Suppl 591:80–87

Stahl B, Chou JH, Li C, Südhof TC, Jahn R (1995) Rab3 reversibly recruits rabphilin to synaptic vesicles by a mechanism analogous to raf recruitment by ras. EMBO J 15:1799–1809

Star RA, Nonoguchi H, Balaban R, Knepper MA (1988) Calcium and cyclic adenosine monophosphate as second messengers for vasopressin in the rat inner medullary collecting duct. J Clin Invest 81:1879–1888

Stetson DL, Lewis SA, Alles W, Wade JB (1982) Evaluation by capacitance measurements of antidiuretic hormone induced membrane area changes in toad bladder. Biochim Biophys Acta 689:267–274

Stow JL, de Almeida JB, Narula N, Holtzman EJ, Ercolani L, Ausiello DA (1991) A heterotrimeric G protein, G alpha i-3, on Golgi membranes regulates the secretion of a heparan sulfate proteoglycan in LLC-PK1 epithelial cells. J Cell Biol 114:1113–1124

Stowell MHB, Marks B, Wigge P, McMahon HAT (1999) Nucleotide-dependent conformational changes in dynamin:evidence for a mechanochemical molecular spring. Nature Cell Biol 1:27–32

Strange K, Willingham MC, Handler JS, Harris, Jr, HW (1988) Apical membrane endocytosis via coated pits is stimulated by removal of antidiuretic hormone from isolated, perfused rabbit cortical collecting tubule. J Membr Biol 103:17–28

Südhof T (1995) The synaptic vesicle cycle:a cascade of protein-protein interactions. Nature 375:645–653

Sutton RB, Fassauer D, Jahn R, Brunger AT (1998) Crystal structure of a SNARE complex involved in synaptic exocytosis at a 2.4 A resolution. Nature 395:347–353

Sweet D (1999) The SAC, EEA1, rab5 and endosome fusion. Trends Cell Biol 9:149

Takacs-Jarrett M, Sweeney WE, Avner ED, Cotton CU (1998) Morphological and functional characterization of a conditionally immortalized collecting tubule cell line. Am J Physiol 275:F802–F811

Takahashi N, Kadowaki T, Yazaki Y, Ellis-Davies GCR, Miyashita Y, Kasai H (1999) Post-priming actions of ATP on Ca^{2+}-dependent exocytosis in pancreatic beta cells. Proc Natl Acad Sci USA 96:760–765

Takei K, Slepnev VI, Haucke V, De Camilli P (1999) Functional partnership between amphiphysin and dynamin in clathrin-mediated endocytosis. Nature Cell Biol 1:33–39

Tamarappoo BK, Verkman AS (1998) Defective aquaporin-2 trafficking in nephrogenic diabetes insipidus and correction by chemical chaperones. J Clin Invest 101:2257–2267

Tasken K, Skalhegg BS, Tasken KA, Solberg R, Knutsen HK, Levy FO, Sandberg M, Orstavik S, Larsen T, Johansen AK, Vang T, Schrader HP, Reinton NT, Torgersen KM, Hansson V, Jahnsen T (1997) Structure, function, and regulation of human cAMP-dependent protein kinases. Adv Second Messenger Phosphoprotein Res 31:191–204

Taylor A, Mamelak M, Reaven E, Maffly R (1973) Vasopressin:possible role of microtubules and microfilaments in its action. Science 181:347–350

Taylor SS, Buechler JA, Yonemoto W (1990) cAMP-dependent protein kinase: framework for a diverse family of regulatory enzymes. Ann Rev Biochem 59:971–1005

Teitelbaum I (1991) Vasopressin-stimulated phosphoinositide hydrolysis in cultured rat inner medullary collecting duct cells is mediated by the oxytocin receptor. J Clin Invest 87:2122–2126

Teitelbaum L, McGuinnes S (1995) Vasopressin resistance in chronic renal failure. Evidence for the role of decreased V2 receptor mRNA. J Clin Invest 96:378–385

Terris J, Ecelbarger CA, Marples D, Knepper MA, Nielsen S (1995) Distribution of aquaporin-4 water channel expression within the rat kidney. Am J Physiol 269:F775–F785

Tsukita S, Yonemura S, Tsukita S (1997) ERM proteins:head to tail regulation of actin-plasma membrane interaction. TIBS 22:53–58

Tuvim MJ, Adachi R, Chocano JF, Moore RH, Lampert RM, Zera E, Romero E, Knoll BJ, Dickey BF (1999) Rab3D, a small GTPase, is localized on mast cell secretory granules and translocates to the plasma membrane upon exocytosis. Am J Respir Cell Mol Biol 20:79–89

Ungermann C, Nichols BJ, Pelham HRB, Wickner W (1998a) A vacuolar v-t-SNARE complex, the predominant form *in vivo* and on isolated organelles, is disassembled and activated for docking and fusion. J Cell Biol 140:61–69

Ungermann C, Sato K, Wickner W (1998b) Defining the function of trans-SNARE pairs. Nature 396:543–548

Vajanaphanich M, Schultz C, Tsien RY, Traynor-Kaplan AE, Pandol SJ, Barret KE (1995) Cross-talk between calcium and cAMP-dependent intracellular signaling pathways. J Clin Invest 96:386–393

Valenti G, Frigeri A, Ronco PM, D'Ettorre C, Svelto M (1996) Expression and functional analysis of water channels in a stably AQP2-transfected human collecting duct cell line. J Biol Chem 271:24365–24370; Correction (1997) J Biol Chem 272:26794

Valenti G, Procino G, Liebenhoff U, Frigeri A, Benedetti PA, Ahnert-Hilger G, Nürnberg B, Svelto M, Rosenthal W (1998) A heterotrimeric G protein of the G_i family is required for cAMP-triggered trafficking of aquaporin 2 in kidney epithelial cells. J Biol Chem 273:22627–22634

Valentijn JA, Gumkowski FD, Jamieson JD (1996) The expression pattern of rab3D in the developing rat exocrine pancreas coincides with the acquisition of regulated exocytosis. Eur J Cell Biol 71:129–136

Valtin H (1976) Animal model of human disease:hereditary hypothalamic diabetes insipidus. Am J Pathol 83:633–636

Valtin H (1992) Genetic models of diabetes insipidus. In:Handbook of Physiology. Renal Physiology. Bethesda, MD. Am Physiol Soc sect 8 vol 2:1281–1315

Van Aelst L, D'Souza-Schorey C. (1997) Rho GTPases and signaling networks. Genes Dev 11:2295–322

Vaughan M (1998) Signaling by heterotrimeric G proteins minireview series. J Biol Chem 273:667–668

Verkman AS, Lencer WI, Brown D, and Ausiello DA (1988) Endosomes from kidney collecting tubule cells contain the vasopressin-sensitive water channel. Nature 333:268–269

Verkman AS, Weyer P, Brown D, Ausiello DA (1989) Functional water channels are present in clathrin-coated vesicles from bovine kidney but not from brain. J Biol Chem 264:20608–20613

Verkman AS (1992) Water channels in cell membranes. Annu Rev Physiol 54:97-108

Vijayaraghavan S, Goueli SA, Davey MP, Carr DW (1997) Protein kinase A-anchoring inhibitor peptides arrest mammalian sperm motility. J Biol Chem 272:4747–4752

Vischer UM, Wollheim CB (1998) Purine nucleotides induce regulated secretion of von Willebrand factor:involvement of cytosolic Ca^{2+} and cyclic adenosine monophosphate-dependent signalling in endothelial exocytosis. Blood 91:118–127

Vischer UM, Lang U, Wollheim CB (1998) Autocrine regulation of endothelial exocytosis:von Willebrand factor release is induced by prostacylin in cultured endothelial cells. FEBS Lett 424:211–215
Vitale N, Mukai H, Rouot B, Thiersé D, Aunis D, Bader M-F (1993) Exocytosis in
    chromaffin cells. J Biol Chem 268:14715–14723
Von Gersdorff H, Matthews G (1999) Electrophysiology of synaptic vesicle cycling.
    Annu Rev Phsyiol 61:725–52
Wade JB, Stetson DL, Lewis SA (1981) ADH action:evidence for membrane shuttle
    mechanism. Ann NY Acad Sci 372:106–117
Wade JB (1994) Role of membrane traffic in water and Na responses to vasopressin.
    Semin Nephrol. 14 :322–332
Wagner C, Kurtz A (1998) Regulation of renin release. Curr Opin Nephrol Hypertens
    7:437–441
Wang G, Witkin JW, Hao G, Bankaitis VA, Scherer PE, Baldini G (1997) Syndet is a
    novel SNAP-25 related protein expressed in many tissues. J Cell Sci 110:505–513
Wang Y, Okamoto M, Schmitz F, Hofmann K, Südhof TC (1997) RIM is a putative
    Rab3 effector in regulating synaptic-vesicle fusion. Nature 388:593–598
Weber E, Berta G, Tousson A, St John P, Green MW, Gopalokrishnan U, Jilling T,
    Sorscher EJ, Elton TS, Abrahamson DR, Kirk KL (1994) Expression and polarized
    targeting of a rab3 isoform in epithelial cells. J Cell Biol 125:583–594
Weber T, Zemelman BV, McNew JA, Westermann B, Gmachl M, Parlati F, Söllner
    TH, Rothman JE (1998) SNAREpins:minimal machinery for membrane fusion.
    Cell 92:759–772
Wollheim CB, Lang J, Regazzi R (1996) The exocytotic process of insulin secretion
    and its regulation by Ca^{2+} and G proteins. Diabetes Rev 4:276–297
Wonerow P, Schöneberg T, Schultz G, Gudermann T, Paschke R (1998) Deletions in
    the third intracellular loop of the thyrotropin receptor. A new mechanism for
    constitutive activation. J Biol Chem 273:7900–7905
Xu DL, Martin PY, Ohara M, StJohn J, Pattison T, Meng X, Morris K, Kim JK, Schrier
    RW (1997) Upregulation of aquaporin-2 water channel expression in chronic
    heart failiure rat. J Clin Invest 99:1500–1505
Yamaguchi T, Yamamoto A, Furuno A, Hatsuzawa K, Tani K, Himeno M, Tagaya M
    (1997) Possible involvement of heterotrimeric G proteins in the organization of
    the Golgi apparatus. J Biol Chem 272:25260–25266
Yamaki M, Kusano E, Tetsuka T, Takeda S, Homma S, Murayama N, Asano Y (1991)
    Cellular mechanisms of lithium-induced nephrogenic diabetes insipidus in rats.
    Am J Physiol 261:F505–F511
Yamaki M, McIntyre S, Rassier ME, Schwartz JH, Dousa TP (1992) Cyclic-3',5'-
    nucleotide diesterases in dynamics of cAMP and cGMP in rat collecting duct
    cells. Am J Physiol 262:F957–F964
Yang B, Gonzalez Jr L, Prekeris R, Steegmaier M, Advani RJ, Scheller RH (1999)
    SNARE interactions are not selective. J Biol Chem 274:5649–5653
Yasui M, Zelenin SM, Celsi G, Aperia A (1997) Adenylate cyclase-coupled vasopressin receptor activated AQP2 promoter via a dual effect on CRE and AP1 elements. Am J Physiol 272:F443–F450
Yasui M, Kwon TH, Knepper MA, Nielsen S, Agre P (1999) Aquaporin-6:An intracellular vesicle water channel protein in renal epithelia. Proc Natl Acad Sci USA
    96:5808–5813

Zeidel ML, Hammond T, Bothelo B, Harris Jr HW (1992) Functional and structural characterization of endosomes from toad bladder epithelial cells. Am J Physiol 263:F62–F76

Zhu X, Gilbert S, Birnbaumer M, Birnbaumer L (1994) Dual signaling potential is common among Gs-coupled receptors and dependent on receptor density. Mol Pharmacol:460–469

Zussman A, Hermuet S, Sagi-Eisenberg R (1998) Stimulation of Ca^{2+}-dependent exocytosis and arachidonic acid release in cultured mast cells (RBL-2H3) by a GTPase-deficient mutant of G alpha i3. Eur J Biochem 258:144–149

Molecular Water Pumps

T. Zeuthen

The Panum Institute, Blegdamsvej 3C, DK 2200 Copenhagen N, Denmark

Contents

1
Introduction

Molecular water pumps are membrane proteins in which a flux of water is coupled to the fluxes of the non-aqueous substrates (Fig. 1). Energy is exchanged between the fluxes; the free energy stored in the gradients of the non-aqueous substrates is coupled, within the protein, to the transport of water. Accordingly, the flux of water is relatively independent of its external chemical potentials and may even proceed uphill. In short, water is being pumped.

The concept has its origin in two experimental findings in epithelial cells from gall-bladder. First, the cells retained their intracellular osmolarity when adapted to steady states in dilute external solutions. Second, osmotically induced in- or effluxes of water across the cell membrane induced a simultaneous flux of K^+. Intracellular hyperosmolarities as high as 40 mosm l^{-1} were observed, and the ratio of the K^+ fluxes to water fluxes was constant about 100 mmol l^{-1} (Zeuthen 1981, 1982, 1983). The experiments raised the fundamental question: How can a cell be hyperosmolar relative to the surroundings in a steady state? Two conclusions were reached: First, the coupling between K^+ and water was a property of a membrane protein. Second, this coupling was of a magnitude that allowed cellular water homeostasis to be achieved as a steady state between an osmotic influx of water and an efflux coupled to the ubiquitous K^+ leak.

The application of volume sensitive microelectrodes (Reuss 1985) to epithelial cells demonstrated that coupling between fluxes of K^+ and of water can take place in K^+/Cl^- cotransport proteins and that the coupling has the

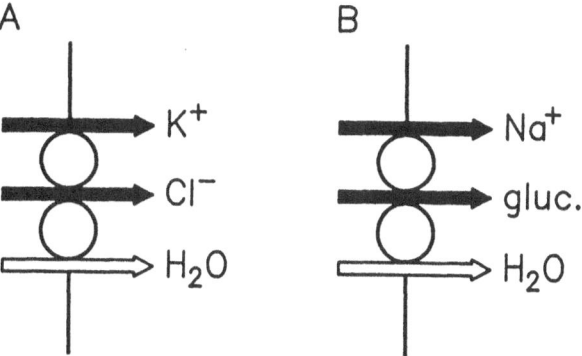

Fig. 1. Molecular water pumps are membrane proteins in which a flux of water is coupled to the substrate fluxes. Two examples are shown, the electroneutral K^+/Cl^- cotransporter and the electrogenic $Na^+/glucose$ cotransporter

character of cotransport (Zeuthen 1991a, 1991b, 1994). Similar experiments on the H^+/lactate cotransporter (Zeuthen et al. 1996) and the Na^+/alanine cotransporter (Zeuthen 1992) also demonstrated coupled water transport.

The introduction of the *Xenopus* expression system for the study of cloned cotransporters (Hediger et al. 1987) has recently allowed a number of cotransporters, in particular Na^+ dependent and electrogenic ones, to be identified as molecular water pumps. The combination of voltage clamp with fast and precise recordings of water transport rates has led to accurate definitions of the coupling properties, and to the experimental exclusion of possible artifacts such as unstirred layers (Loo et al. 1996, Zeuthen et al. 1997, Meinild et al. 1998a). It seems that coupled water transport is a general property of cotransporters of the symport type (Table 1).

The properties of the molecular water pumps may derive from cotransport of water, in which case water is a substrate on equal footing with the non-aqueous substrates. On this model a downhill flux of any of the substrates, also water, can drive the uphill flux of the others in a fixed stoichiometrical relationship given by the protein. We may talk about, for example, K^+/Cl^-/H_2O cotransport. In a less strict mode of coupling, secondary active transport of water, the downhill fluxes of the non-aqueous substrates lead to the uphill transport of water, while the opposite may not be the case: a downhill flux of water may not lead to uphill fluxes of the other substrates. Two molecular models are discussed in Section 7. One, the mobile barrier model, gives rise to cotransport of water. The other, the osmotic coupling model, results in secondary active transport of water. The precise molecular mechanism behind the molecular water pumps is as yet unknown. However, phenomena well described for aqueous enzymes undergoing conformational changes do point towards the cotransport type of model.

The energy used by the molecular water pumps derives ultimately from trans-membrane differences in chemical and electrochemical potentials. These, in turn, arise from transport in primary active transport proteins, such as the Na^+/K^+ATPase, or they can be imposed by the experimental conditions. The introduction of molecular water pumps as new building blocks in cellular models will, in combination with the primary active transport proteins and channels, explain a number of experimental findings, both in regard to cellular water homeostasis and to trans-epithelial water transport.

Table 1. Number of water molecules (n) transported per turnover in cotransport proteins

Cotransporter Organ Species	Number of water molecules, n.	Method	Reference
K^+/Cl^- Choroid plexus *Necturus maculosus*	500	Microelectrodes	Zeuthen 1994
H^+/lactate Retina *Necturus maculosus* Human, cultures	500 Not determined	Microelectrodes Fluorescence	Zeuthen et al. 1996, Hamann et al. 1996
Na^+/alanine Small intestine *Necturus maculosus*	Not determined	Microelectrodes	Zeuthen 1995
Na^+/glucose, SGLT1 Rabbit Human	390 210	Expression in *Xenopus* oocytes	Loo et al.1996, Zeuthen et al. 1997, Meinild et al. 1998a
Na^+/I^- Rat	200	Expression in *Xenopus* oocytes	Loo et al. 1996
Na^+/Cl^-/GABA Rat	~200	Expression in *Xenopus* oocytes	Loo et al. 1996
H^+/amino acid Plant	50	Expression in *Xenopus* oocytes	Loo et al. 1996
Na^+/dicarboxylate Rabbit	175	Expression in *Xenopus* oocytes	Meinild et al. 2000
Na^+/glutamate Human	425	Expression in *Xenopus* oocytes	Boxenbaum, personal communication

2
Uphill Transport of Water in Cotransport Proteins

Uphill transport of water across biological membranes has been recognized for more than a hundred years (Ludwig 1861 (pg. 346), Reid 1892 and 1901). In leaky epithelia (i.e. kidney proximal tubule, small intestine, choroid plexus) water can be transported across the cell layer in the direction of higher water chemical potentials. The small intestine can transport against 200 mosm l^{-1} equivalent to a hydrostatic pressure difference of about 40 meters of water, for references see Zeuthen (1992, 1995, 1996) and Zeuthen and Stein (1994). It is generally accepted that the energy needed for the water transport across epithelia is derived from differences in electrochemical potentials for ions. While the links between active ion transport and ionic gradients are well established, the coupling between the ionic gradients and the transport of water is not understood at all. The possibility examined here is that the coupling between water fluxes and ionic gradients can take place directly in membrane proteins of the cotransport type.

This section discusses some key experiments that demonstrate coupling between the fluxes of non-aqueous substrates and water. In particular, experiments where downhill fluxes of non-aqueous substrates drive uphill fluxes of water. The evidence that unstirred layers do not interfere with the water fluxes is discussed in Section 3. Experiments aimed at a more precise definition of the coupling mechanism are reviewed in Section 4.

2.1
The K^+/Cl^- Cotransporter

The interaction between ions and water in the K^+/Cl^- cotransporter is most conveniently studied in the epithelium from choroid plexus, Fig. 2. The epithelium produces cerebrospinal fluid by transporting ions and water from the blood plasma into the brain ventricles. It is a so-called backward epithelium since the membrane proteins are distributed among the apical and serosal membranes oppositely to what is found for most other epithelia. This gives the experimental advantage that the K^+/Cl^- cotransporter is located in the exposed apical membrane, where it can be studied by means of ion selective microelectrodes under short-lasting abrupt shifts in bathing solutions composition (Zeuthen 1991a, 1991b, 1994). Volume changes of the epithelial cell were recorded by means of a volume sensitive microelectrode (Reuss 1985): Choline chloride (1 mmol l^{-1}) was added to the bathing solutions and the cellular concentration of choline ions (Ch^+) became relative stable (< 1 mmol l^{-1}) within 30 minutes. Because K^+ sensitive microelectrodes are much more sensitive to Ch^+ than to K^+, cellular volume changes can be recorded as changes in Ch^+ concentrations.

The osmolarity of the cerebrospinal fluid compartment was first increased abruptly by 100 mosm l^{-1} by means of mannitol or NaCl; this caused the cell to shrink as expected. If, however, the osmolarity was increased

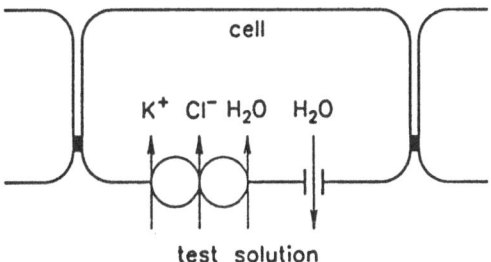

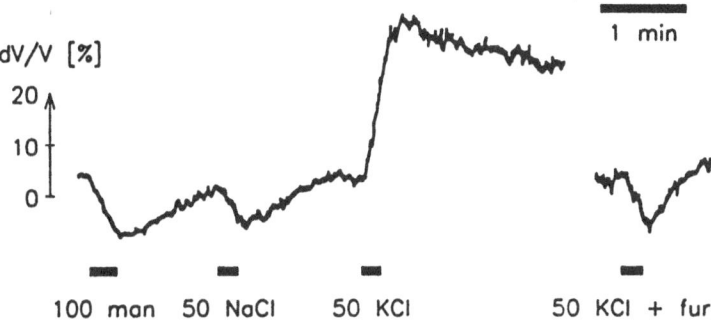

Fig. 2. Uphill transport of water by the K^+/Cl^- cotransporter. The apical membrane of choroid plexus epithelium from *Necturus maculosus* was exposed to hypertonic test solutions. The epithelial cell volume was recorded by means of a volume sensitive microelectrode. Addition of 100 mmol l^{-1} of mannitol or 50 mmol l^{-1} of NaCl caused the cell to shrink. Addition of 50 mmol l^{-1} of KCl caused the cell to swell despite the fact that the external osmolarity was now higher than the intracellular osmolarity by 100 mosm l^{-1}. The intracellular concentrations of K^+ and Cl^- only changed a few mmol l^{-1} during the test. Apparently, the KCl induces an uphill influx of water which is much larger than the water which leaves the cell via the passive pathways (insert). If the cotransporter was blocked by 10^{-4} mol l^{-1} of furosemide (fur) the cell shrank in response to the addition of KCl. Adapted from Zeuthen (1994)

100 mosm l^{-1} by the addition of KCl the cell immediately began to swell. If only passive osmotic transport was operating in the membrane, this solute gradient should have caused cell shrinkage in analogy to the mannitol and NaCl experiments. Recordings of the intracellular ion activities (Na^+, K^+, and Cl^-) showed that the swelling took place before the intracellular osmolarity had changed significantly. Consequently, the direction of water movement is opposite to that given by the difference in water chemical potential across the membrane. The the cell swelling is an example of uphill

transport of water. In the presence of furosemide (10^{-3} mol l^{-1}) or bumetanide (10^{-4} mol l^{-1}), addition of KCl caused cell shrinkages similar to
those obtained by mannitol or NaCl.

2.2
The H$^+$/Lactate Cotransporter

Water transport by the H$^+$/lactate cotransporter was studied in the retinal
membrane of the retinal pigment epithelium of the eye, Fig. 3. This epithe-

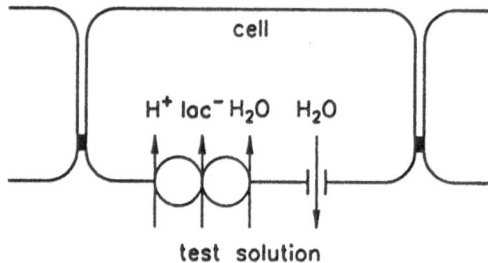

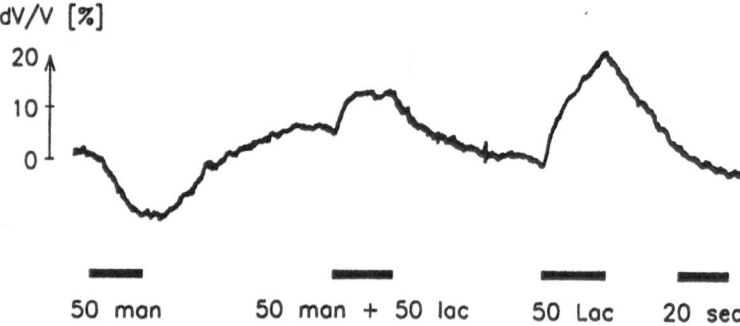

Fig. 3. Uphill transport of water by the H$^+$/lactate cotransporter. The retinal surface
of epithelial cells from the retinal pigment epithelium of Bullfrog were exposed
abruptly to various test solutions. The cell volume was recorded by means of a volume sensitive microelectrode. As expected, addition of 50 mmol l^{-1} of mannitol
caused the cell to shrink (50 man). When 50 mmol l^{-1} of Cl$^-$ was replaced isosmotically by lactate ions the cells swelled abruptly (50 lac). When mannitol was added
simultaneously with the lactate replacement, the cells swelled (50 man + 50 lac). In
this situation water moves into the cell against an osmotic difference of 50 mosm l^{-1}.
Apparently, the influx of lactid acid via the cotransporter induces an uphill influx of
water which is much larger than the water which leaves the cell via the passive pathways (insert). The H$^+$ and lactate concentrations inside the cell did not change significantly during the test. Adapted from Zeuthen et al. (1996)

lium is located directly under the neuroretina and serves to remove water and metabolites from the retinal space and into the chroidal blood supply. In vivo this water absorbtion is held important for the maintenance of normal retinal adhesion (Marmor 1989). The retinal surface of the epithelium can be exposed by removing the neuro-retina by disection. We have employed both ion and volume sensitive microelectrodes (Zeuthen et al. 1996) as well as fluorescence methods (Hamann et al. 1996).

When mannitol was added to the retinal bathing solution the cell shrank as would be expected from normal osmometric behaviour, this experiment also gave the osmotic water permeability, L_p, of the retinal membrane. If Cl⁻ ions were replaced isotonically by lactate ions the H^+/lactate transporter was activated and an abrupt cell swelling observed. If the two changes were performed simultaneously, i.e. addition of manitol and activation by lactate, the cell swelled. In this experiment water moved into the cell in the direction from a high to a low osmolarity. The change in intracellular osmolarity estimated from the influx of H^+ and lactate was much too slow to increase the intracellular osmolarity significantly during the test. If the lactate induced swellings should be explained by simple osmosis it would require the sudden emergence of 200 mosm l^{-1} at the intracellular surface of the membrane in order to pull water in at the observed rate. This is unlikely, the influxes of H^+ and lactate only increased the intracellular osmolarity at a rate of 0.25 mosm l^{-1} sec^{-1}. We concluded that the downhill influxes of H^+ and lactate ions in the cotransporter energized an uphill influx of water.

2.3
The Na⁺/Glucose Cotransporter

The Na⁺/glucose cotransporter from mammalian small intestine (SGLT1) has been cloned (Hediger et al. 1987) and expressed in *Xenopus* oocytes (Parent et al. 1992). The transporter is electrogenic, it transports two Na⁺ ions for each glucose molecule. Accordingly, the influx can be monitored as an inward current under clamp conditions (Mackenzie et al. 1998). The round shape of the oocyte allows the fluxes of water to be determined from changes in the cross section of the oocyte monitored by an optical method (Loo et al. 1996, Zeuthen et al. 1997). The bathing solution could be changed in a few seconds and the relative volume changes recorded with an accuracy of 0.03%.

When glucose replaced mannitol isotonically in the bathing solution and the intracellular electrical potential was clamped to a negative value, we observed an abrupt increase in inward current and a linear increase in oocyte

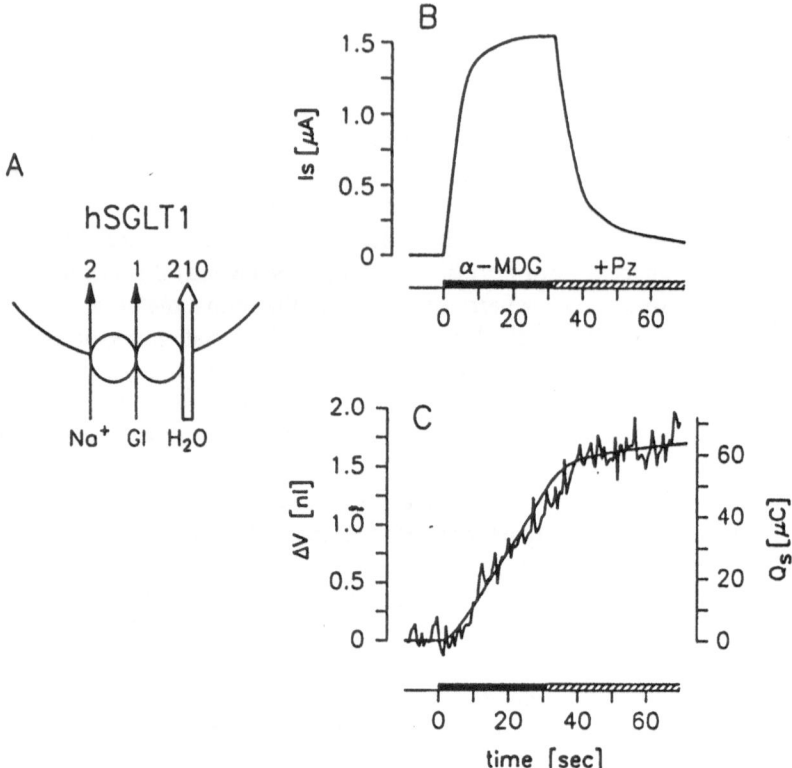

Fig. 4. Correspondence between sugar induced current and water flow in the Na⁺/glucose transporter under isotonic conditions. A. The Na⁺/glucose cotransporter from the human small intestine (hSGLT1) was expressed in *Xenopus* oocytes. The oocyte was clamped at a membrane potential of -50mV by two-microelectrode voltage clamp. B. 10 mmol l⁻¹ of a glucose-analog, methyl-α-D-glucopyranoside (α-MDG), was introduced abruptly into the bath (black bar) where it replaced, isotonically, 10 mmol l⁻¹ of mannitol. This caused an inward current (I_s), which increased rapidly towards a maximum of 1540 nA (90% complete in 13 sec). C. The current was associated with a linear increase in oocyte volume (ΔV, jagged line) indicative of a constant rate of water influx of 36 pl sec⁻¹. The volume increase and the current were abolished by the addition of 50 μmol l⁻¹ of phlorizin (+Pz). The integrated current (Q_s, smooth line in C), describes the volume changes if we assume that the entry of 2 Na⁺ ions and 1 sugar molecule couples directly to the entry of 249 water molecules. The average coupling ratio of all experiments was 210 water molecules per 2 Na⁺ and 1 sugar molecule. Adapted from Meinild et al. (1998a)

volume, Fig. 4. The correspondence between the glucose induced current and the inflow of water was shown by the fact that the integrated current closely predicted the change in volume if it was assumed that a fixed number of water molecules was coupled to the influx of each Na^+ ion. Our resent estimates are 390 water molecules per two Na^+ and one glucose for the rabbit SGLT1 (Zeuthen et al. 1997) and 210 water molecules for the human SGLT1 (Meinild et al. 1998a).

The ability for uphill water transport in the Na^+/glucose transporter was investigated in details in the human SGLT1, Fig. 5A and B. If the bathing solution was made hypertonic under clamp conditions by the addition of 10 mmol l^{-1} glucose, the induced short circuit current (I_s) was associated with an influx of water. The water influx is a sum of two fluxes: a sugar coupled inward flux and an outward osmotic flux. The osmotic component could be obtained by adding 10 mmol l^{-1} mannitol instead of glucose, Fig. 5A. The glucose induced component is then obtained by subtraction, Fig. 5B. This component of the water flux was closely linked to the short circuit current. If we assume that a fixed number of water molecules are transported for each cycle of the protein then the integrated current corresponds to the glucose induced component of the water transport. The coupling ratio was the same as that observed under isotonic conditions, 210 water molecules were transported for each two Na^+ and one glucose molecule (Meinild et al. 1998a).

It was also possible to divide the water flux of the SGLT1 into a glucose induced and an osmotic component in experiments performed under hypotonic conditions, Fig. 5C and D (Loo et al. 1996, Meinild et al. 1998a). When 10 mmol l^{-1} mannitol was removed from the bathing solution the oocyte swelled. When the same osmotic gradient was implemented by removing 20 mmol l^{-1} mannitol and adding 10 mmol l^{-1} glucose the oocyte swelled at a faster rate, Fig. 5C. The glucose induced component of the water flux could be obtained by subtraction, Fig. 5D. It had the same ratio to the integrated current as that obtained in the isotonic and the hypertonic experiments, for the human SGLT1 210 water molecules per two Na^+ and one sugar molecule. The water fluxes in the human SGLT1 under the different osmotic conditions are summarized in Fig. 18A.

2.4
Other Transporters

So far, all cotransporters of the symport type have exhibited coupled transport of water (Table 1). The list includes cotransporters from both amphibians, mammals and plants, which underlines the generality of the phenom-

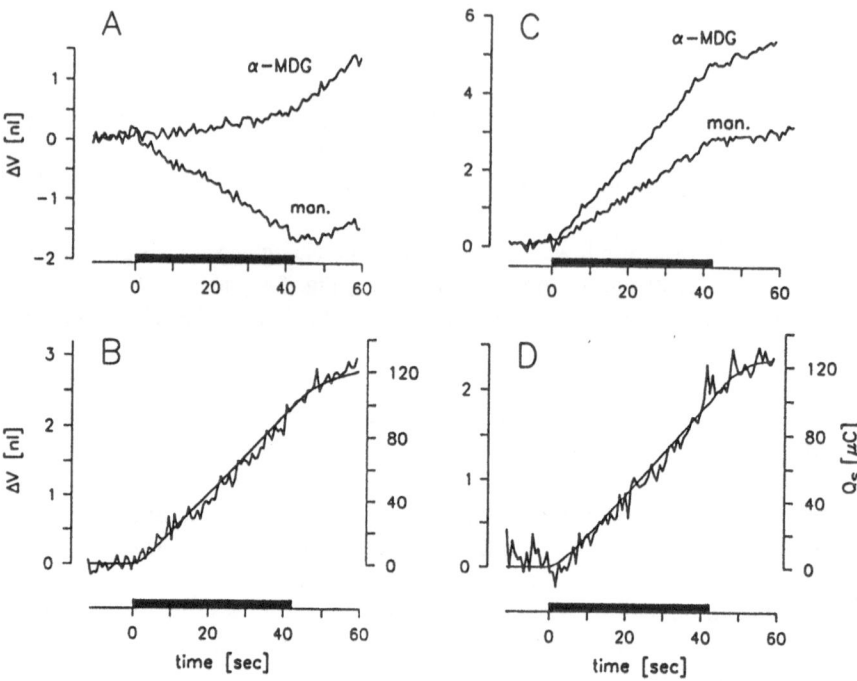

Fig. 5. Uphill (A and B) and downhill (C and D) transport of water by the human Na^+/glucose transporter, hSGLT1. An hSGLT1 expressing oocyte was clamped to -50 mV, see Fig. 4. In **A**, 10 mmol l^{-1} of mannitol was replaced by 20 mmol l^{-1} of a glucose-analog (α-MDG, black bar) equivalent to an increase of the extracellular osmolarity of 10 mosm l^{-1}. This caused an inwardly directed current (not shown) and a linear increase in oocyte volume ΔV (top curve). If instead the extracellular osmolarity was increased 10 mosm l^{-1} by the addition of 10 mmol l^{-1} of mannitol, the oocyte shrank as expected (man). The difference between the two volume changes is given as the jagged line in **B** and represents the uphill transport of water mediated by the downhill fluxes of Na^+ and glucose in the hSGLT1. The integrated current (Q_s, smooth line in **B**) describes the volume changes if it is assumed that about 210 water molecules were transported for each two Na^+ ions and one glucose molecule. In **C**, 20 mmol l^{-1} of mannitol was replaced abruptly by 10 mmol l^{-1} of the glucose-analog (α-MDG, black bar). This caused an inwardly directed current (not shown) and a rapid linear increase in oocyte volume (top curve). If instead the extracellular osmolarity was decreased by 10 mosm l^{-1} by removing mannitol (no sugar present), the oocyte swelled at a slower rate (man.). The difference between the two volume changes represents the transport of water mediated by the hSGLT1 under hypotonic conditions, i.e. downhill transport (**D**, jagged line). The integrated current (Q_s, smooth line in **D**) describes the volume changes if it is assumed that about 210 water molecules are transported together with two Na^+ ions and one glucose molecule. See also Fig. 18A. Adapted from Meinild et al. (1998a)

ena. Interestingly, the coupling ratio (the number of water molecules transported per transport cycle) is largest for the transporters from amphibia, about 500, intermediate for the mammalian group, in the range 175 to 400, while it is small for the cotransporter from plants, about 50. This sequence parallels the plasma osmolarities of the hosts. It is tempting to speculate that the cotransport of water in the different species is tailored so as to minimize the disturbance of the intra and extra cellular osmolarities during transport. The antiporters Na^+/H^+ (Zeuthen, unpublished) and the Cl^-/HCO_3^- (Cotton, and Reuss, 1991) apparently did not have coupled water transport.

Recently a different type of molecular pump has been suggested (Baslow 1998, 1999). N-acetyl-L-histidine (NAH) is assumed to leave the cells of the lens in the eye associated with water, about 33 water molecules per molecules of NAH. Subsequently, NAH is broken down extra-cellularly and the end products recycled without water to the cells. The experimental evidence is as yet circumstantial, there is no direct measurements of the transmembrane osmotic differences or the coupling ratio.

3
Unstirred Layers

Coupled water transport could be explained as an effect of unstirred layers if two conditions were fulfilled: 1. The flows of non-aqueous substrates lead to a build-up or a reduction of concentrations and osmolarities in the vicinity of the membrane and 2) the osmotic water permeability of the membranes is underestimated by one or more orders of magnitude. Under these conditions the difference in water chemical potential across the membrane would be different to that given by the external bulk solution and the cytoplasm. Combined with a high water permeability of the membrane, it might require only minor unstirred layers to drive large, and even uphill, fluxes of water. Fortunately, both calculations as well as direct experimental evidence clearly rule out significant unstirred layer effects.

A straightforward experimental demonstration of the absence of unstirred layer effects has been obtained for the human and the rabbit Na^+/glucose transporters (SGLT1) expressed in the *Xenopus* oocyte Fig. 6, (Zeuthen et al. 1997, Meinild et al. 1998a). The cation-selective ionophore gramicidin was inserted into the membrane of the SGLT1 expressing oocytes under conditions where the cotransporter was kept inactivated by the absence of glucose. Clamping the oocyte to a potential in the range −50 to −150 mV led to inward clamp currents of 750 to 1900 nA most likely carried by Na^+. Importantly, there was no associated change in oocyte volume.

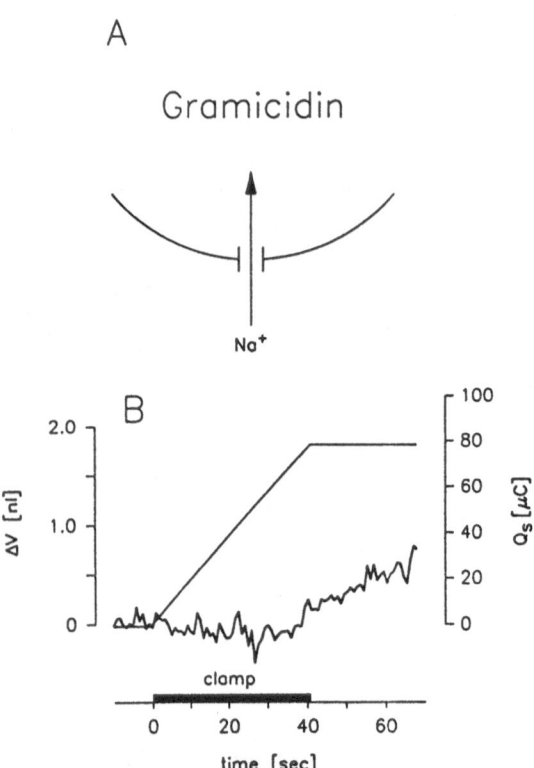

Fig. 6. Demonstration of no unstirred layer effects associated with Na^+ fluxes. **A.** The cation selective ionophore Gramicidin was inserted into oocytes expressing hSGLT1 as in Fig. 4, but the cotransporter was kept inactivated by the absence of glucose. **B.** An inward current of about 1900 nA was obtained (not shown) by changing the intracellular electrical potential from the resting potential, about -45 mV, to -100 mV by means of the voltage clamp (black bar). The current, carried by Na^+ through the Gramicidin, did not give rise to any immediate change in cell volume (jagged line) and the integrated current (smooth line) had no resemblance to the volume changes. Adapted from Meinild et al. (1998a)

Similar results were obtained with the ionophore nystatin (Zeuthen et al. 1997) and the cation channel Connexin 50 (White et al. 1992, Zampighi et al. 1997). Connexin 50 was co-expressed in the SGLT expressing oocytes; in this case clamp currents in the range 500–4000 nA were achieved (Meinild et al. 1998a). In terms of unstirred layers, the situation where the current is channel mediated is analogous to the situation where the current is carried entirely by the glucose-activated SGLT1. The only difference is the molecular mechanism responsible for the Na^+ transport. It can therefore be concluded that it is the molecular mechanisms within the SGLT1 itself that gives rise to the water transport and not unstirred layers or other artefacts such as ion flows from the clamp electrodes. It should be noted that the two situations are governed by the same passive water permeability given by the passive properties of the SGLT1, see Section 6, and the native oocyte membrane; the extra water permeability supplied by the gramicidin is insignificant (Zeuthen et al. 1997). The glucose that enters the oocyte in cotransport with Na^+ will add to the intracellular osmolarity by up to 50% beyond that given by the clamp current. However, in the experiments with Connexin 50, initial osmotic swelling was absent at clamp currents far exceeding those used in the cotransport experiments. This margin makes it unlikely that the glucose transported in the cotransport experiments should cause any significant initial influx of water by simple osmosis. The experimental evidence for no unstirred layer effects in the *Xenopus* oocyte expressing SGLT1 are supported by theoretical evaluations. Calculations estimate unstirred layer effects to be able to account for around 3% of the active water flux (Loo et al. 1996).

The experimental tests for unstirred layer effects, which are available in the *Xenopus*, are not possible in the intact epithelia from the choroid plexus or the retinal pigment epithelium. The data themselves, however, rule out any significant influence from unstirred layers. Consider Fig. 2. Here the addition of 50 mmol l^{-1} KCl to the external solution resulted in cell swellings of about 1.5% sec^{-1} while the addition of equivalent amounts of NaCl or mannitol caused a cell shrinkage of about 0.5% sec^{-1}. If unstirred layers were to explain the cell swellings induced by KCl, this would require the build-up of a total transmembrane osmotic difference of some 300 mosm l^{-1} with the direction for passive water transport being into the cell. In that case the unstirred layer itself should give rise to a difference of 400 mosm l^{-1} for the total difference to become 300 mosm l^{-1}. Unstirred layers of this magnitude are inconceivable for at least three reasons. First and simplest, the concentrations of KCl in the layer would cause the driving force for the K^+/Cl^- cotransporter itself to vanish or even reverse. Second, conventional unstirred layer theory (Dainty and House 1966) relates solute flow (J_s) and volume flow (J_w) by $J_w = J_s * L_p * RT * (\delta^o + \delta^i)/(D_s + c_s(\delta^o + \delta^i) * L_p * RT)$, where L_p is the water permeability, RT is the product of the gas constant and the absolute temperature, δ is the thickness of the unstirred layers on the outside (o) and the inside (i) of the membrane, D_s is the solute diffusion coefficient and c_s is the concentration of the solute. Unstirred layers in the present type of experiments have been determined to be no more than 25–40 μm (Reuss 1985, Persson and Spring 1982, Cotton and Reuss 1989, Cotton et al. 1989). With L_ps of $0.6*10^{-4}$ cm sec^{-1} (osm l^{-1})$^{-1}$, J_s of $2*10^{-9}$ mol cm^{-2} sec^{-1} (Zeuthen 1994), a

c_s of, say, 200 mmol l^{-1} and a D_s of $4*10^{-6}$ cm^2 sec^{-1} (a low estimate), J_w calculates to be less than 1% of the observed volume flux, $17*10^{-6}$ cm sec^{-1}. Third, unstirred layers arising in a spherical fashion around the orifices of the transporting proteins are unlikely. The turnover number of the transporter is likely to be smaller than, say, 10^3 sec^{-1}, which is to low to result in fluxes large enough to create gradients in the vicinity of the proteins. Similar arguments can be advanced in relation to the measurements on the H^+/lactate cotransporter, see above.

Unstirred layer effects have been held important, but mostly the arguments have been qualitative (Diamond 1979, Barry and Diamond 1984, Spring 1998). In particular, it has been argued that unstirred layers might cause the measured water permeabilities to be orders of magnitude smaller that the true ones. Supposedly, the induced water fluxes sweep the solutes away from the membrane and the truly imposed osmotic gradient would be smaller than the one inferred from the composition of the bulk solution and the cell cytoplasm. Fortunately, when the effects are assessed quantitatively, the corrections are found to be small, no higher than a factor of two (Pedley 1978, 1983, Hill 1980, Pedley and Fischbarg 1980, Persson and Spring 1982, Reuss 1985, Cotton and Reuss 1989, Cotton et al 1989, Boulpaep et al. 1993). The values for the water permeabilities are similar whether obtained with microelectrodes, optical methods, or from experiments on membrane vesicles, see Table 3.3 in Zeuthen (1996). The fact that the same water permeabilities are obtained from intact cells and from vesicle preparations shows clearly that invaginations or microvilli do not lead to unstirred layer effects.

It must be concluded that there is no evidence for unstirred layers in the solutions abutting the protein. A more productive role of unstirred layers will be introduced in Section 7.2 in connection with the discussion of possible molecular mechanisms of coupling between water and solute fluxes. Here the possibility of what might be called intra-molecular unstirred layers will be addressed.

4
Evidence for Cotransport of Water

In cotransport of water, a well-defined number of water molecules act as substrates on equal footing with the non-aqueous substrates. Accordingly, cotransport of water can be recognized experimentally by a number of properties. Not only should the downhill fluxes of non-aqueous substrates be able to drive uphill fluxes of water, as reviewed in Section 2, but downhill fluxes of water should also be able to drive uphill fluxes of the non-aqueous substrates. Furthermore, the ratio of any two of the fluxes should be constant and given by the properties of the cotransport protein alone, i.e. the ratio is independent of external parameters such as driving forces. The proteinaceous character of the cotransport should be manifest by saturation of the fluxes at increasing driving forces, by specificity for the substrates, and by inhibition by specific drugs.

The present section reviews a large body of experimental findings, including those of Section 2, in order to see if they conform to the cotransport mode of coupling. Data on the K^+/Cl^- cotransporter are from the choroid plexus epithelium from *Necturus* (Zeuthen 1991a, 1991b, 1994), data on the H^+/lactate cotransporter are from the retinal pigment epithelium from Bullfrog (Zeuthen *et al.* 1996), and data on the Na^+/glucose transporter are primarily from the human SGLT1 expressed in *Xenopus* oocytes (Meinild *et al.* 1998a).

4.1
Transfer of Energy

In a complete study of a given cotransporter it must be shown that the driving force imposed by one substrate can drive fluxes of all the other substrates. With three substrates this would take six types of experiments, twelve if both influxes and effluxes were to be investigated. The properties of the cell or expression system limit the types of experiments that can be done. Some driving forces have a direction and magnitude that would be lethal to the cells. Clearly, the experimental situation is more versatile the closer the cotransporter is poised towards equilibrium. For these reasons, more types of experiments have been possible with the electroneutral K^+/Cl^- and H^+/lactate cotransporters than with the electrogenic Na^+/glucose cotransporter.

4.1.1
The K^+/Cl^- Cotransporter

All six combinations of downhill and uphill fluxes have been studied: (i) A downhill water flux can drive K^+ and Cl^- fluxes, Fig. 7, (ii) a downhill K^+ flux can drive fluxes of Cl^- and water, Fig. 8, and (iii) a downhill Cl^- flux can drive K^+ and water fluxes. A few comments are warranted in the present context, otherwise the reader is referred to the original papers.

Ad (i). The first evidences of coupling between K^+ and water fluxes were obtained in experiments designed to determine the osmotic water permeability, L_p, of the cell membranes of gall-bladder (Zeuthen 1981, 1982, 1983).

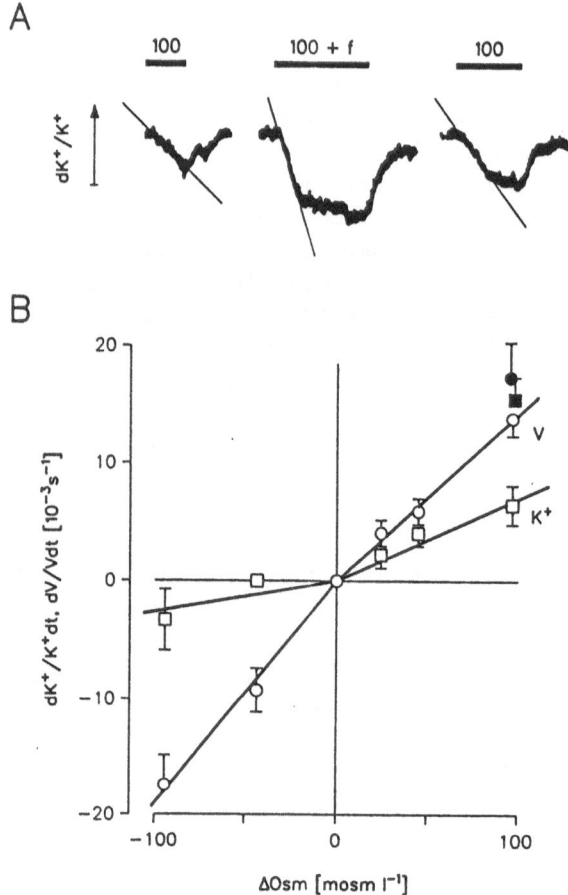

Fig. 7. Water fluxes induces K⁺ fluxes. **A.** Epithelial cells from the choroid plexus
from *Necturus maculosus* were exposed abruptly to solution made hyperosmolar by
addition of 100 mosm l⁻¹ of mannitol and the relative change in intracellular K⁺ concentration was recorded, see Fig. 2. At the bar marked 100 + fur the effect of the simultaneous addition of 10^{-4} mol l⁻¹ of furosemide was tested. There was no simultaneous changes in the intracellular electrical potentials (not shown). **B.** Comparison
between the initial relative rate of change in cell volume V (O) and in intracellular
K⁺ concentration ( ) as a function of the osmotic challenge. Increases in osmolarities
were obtained by adding mannitol, reductions by removing NaCl. The relative rate
of change in K⁺ concentration (dK^+/K^+dt) was always smaller than the corresponding relative rate of change in cell volume (dV/Vdt). This shows that K⁺ moves across
the membrane tightly coupled to the osmotically induced water fluxes. The K⁺ fluxes
were prevented if furosemide was applied to the ventricular membrane; in this case
the relative changes in K⁺ concentration equalled those of the cell volume, filled
symbols. Adapted from Zeuthen (1991a)

Intracellular ion activities were recorded during abrupt changes in extracellular osmolarities and their rates of change were expected to reflect the water permeability of the membrane. This clearly did not hold for the changes in K^+ activities, which were much smaller than those expected from the volume changes. The conclusion was that K^+ was removed from the cell at the same time as the water. The effect was pursued in the choroid plexus epithelial cells, where it was confirmed that osmotically induced effluxes of water caused an immedeate efflux of K^+ while osmotically induced influxes caused immedeate uphill influxes of K^+, Fig. 7. The same conclusions were drawn for the osmotically induced fluxes of Cl^- (Zeuthen 1991a). The coupling between water and ion fluxes was removed by specific blockers of K^+/Cl^- cotransport, furosemide and bumetanide, Fig. 7. The findings implicate a molecular model with water as a substrate, but seem incompatible with an osmotic coupling model, see Section 7.

4.1.2
The H^+/Lactate Cotransporter

Transfer of energy has been demonstrated between any two of the fluxes of H^+, lactate and water (Zeuthen et al. 1996). Specifically, it has been established that: (i) A downhill influx of lactate can drive uphill influxes of water and of H^+, Fig. 3 and 9; (ii) A passive efflux of water can drive an uphill efflux of H^+ if lactate is present. (iii) An inward flux of H^+ accelerates an uphill influx of water induced by lactate. There is no fast method for recording of intracellular lactate concentrations. We have therefore been unable to study the lactate fluxes directly. But as the cotransporter is electroneutral, the lactate flux is assumed equal to the H^+ flux.

Ad (ii) These experiments were performed by adding mannitol to the external solutions while studying the relations between the induced effluxes of water and the changes in intracellular acidity. The ability of the water fluxes to induce uphill fluxes of H^+ is particular important in establishing the underlying molecular mechanism, see Section 7.

4.1.3
The Na^+/Glucose Cotransporter

The types of energy transfer experiments that can be performed on the electrogenic cotransporters expressed in *Xenopus* oocytes are limited. The cotransporters are poised strongly for inward transport *qua* the negative electrical intracellular potential and the Na^+ gradient. In case of the SGLT1 this effect is enhanced by the fact that two Na^+ ions are transported for each transport cycle (Mackenzie et al 1998). For example, osmolarities that might lead to movements of Na⁺ would not be tolerated by the oocytes. Accordingly, we have been restricted to show that downhill fluxes of Na⁺ can drive uphill fluxes of water, Section 2.

4.2
Coupling Ratios

In cotransport the fluxes of the substrates are linked in a fixed stoichiometry. If water is a substrate, it follows that the transported solution has the same composition irrespective of the nature and magnitude of the driving force. The ratio of the fluxes of non-aqueous substrates to that of the water fluxes is called the coupling ratio and given in mmol l^{-1}.

4.2.1
The K⁺/Cl⁻ Cotransporter

The coupling ratio of the K⁺ and Cl⁻ fluxes to the water fluxes was always close to 110 mmol l^{-1} (equivalent to 500 water molecules per K⁺ or Cl⁻), irre-

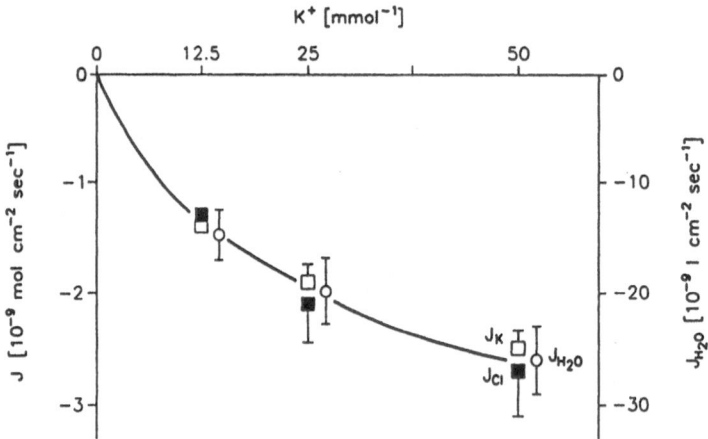

Fig. 8. Coupling ratio and saturation of K⁺, Cl⁻, and water fluxes in the K⁺/Cl⁻ cotransporter. Epithelial cells from the choroid plexus from *Necturus maculosus* were exposed abruptly to test solutions where K⁺ replaced Na⁺ isosmotically, see Fig. 2. The K⁺, Cl⁻, and water fluxes across the ventricular membrane were recorded as initial rates of changes by means of ion and volume sensitive microelectrodes. The ratio of ion to water fluxes was constant irrespective of the driving force; about 110 mmol of K⁺ and Cl⁻ enters per liter of water, equivalent to 500 water molecules per pair of ions. The fluxes exhibited saturation at increasing driving force. Data from Zeuthen (1994)

spective of whether the driving force was derived from a downhill flux of water, K^+, or Cl^-.

The constant coupling ratio is exemplified by experiments where the fluxes of water, Cl^-, and K^+ are driven by downhill fluxes of K^+, Fig. 8. The gradients were established by isotonic replacements of Na^+ by K^+ (Zeuthen 1994). The influxes of K^+ and Cl^- were equal and related to the water fluxes by a constant coupling ratio of 110 mmol l^{-1}. Both ionic fluxes and the water flux were inhibited more than 80% by furosemide.

In experiments where a Cl^- gradient constituted the driving force, the precise value of the Cl^- flux itself was difficult to determine probably due to a flux component via another transporter (Zeuthen 1991b). In these cases the cotransport mediated Cl^- flux could be determined as the furosemide dependent flux.

4.2.2
The H^+/Lactate Cotransporter

The fluxes of H^+, lactate and water in the H^+/lactate cotransporter exhibited coupling ratios with an average of 109 mmol l^{-1}, close to the value found for the K^+/Cl^- cotransporter (Zeuthen et al 1996). This was the case both when downhill lactate or when downhill water fluxes provided the driving force. Experiments where the external H^+ concentration was varied to obtain downhill H^+ fluxes were not feasible. The constancy of the coupling ratio shows that for each H^+ and lactate ion transported about 500 water molecules are cotransported. The fluxes of H^+ and water induced by externally imposed lactate gradients are shown in Fig. 9.

4.2.3
The Na^+/Glucose Cotransporter

In transient experiments in *Xenopus* oocytes, the human Na^+/glucose cotransporter (SGLT1) transported 210 water molecules for each 2 Na^+ and 1 glucose (Meinild et al 1998a). The same coupling ratio was also found in steady states for a range of clamp voltages, external Na^+ concentrations, and external sugar concentrations, see Fig. 10. The coupling ratio was also independent of changes in the external osmolarity, values in the range –20 to 20 mOsm were tested, Fig. 18A. The coupling ratio for the rabbit SGLT1 was higher, our resent estimate is about 390 (Zeuthen etal. 1997), see Table 1.

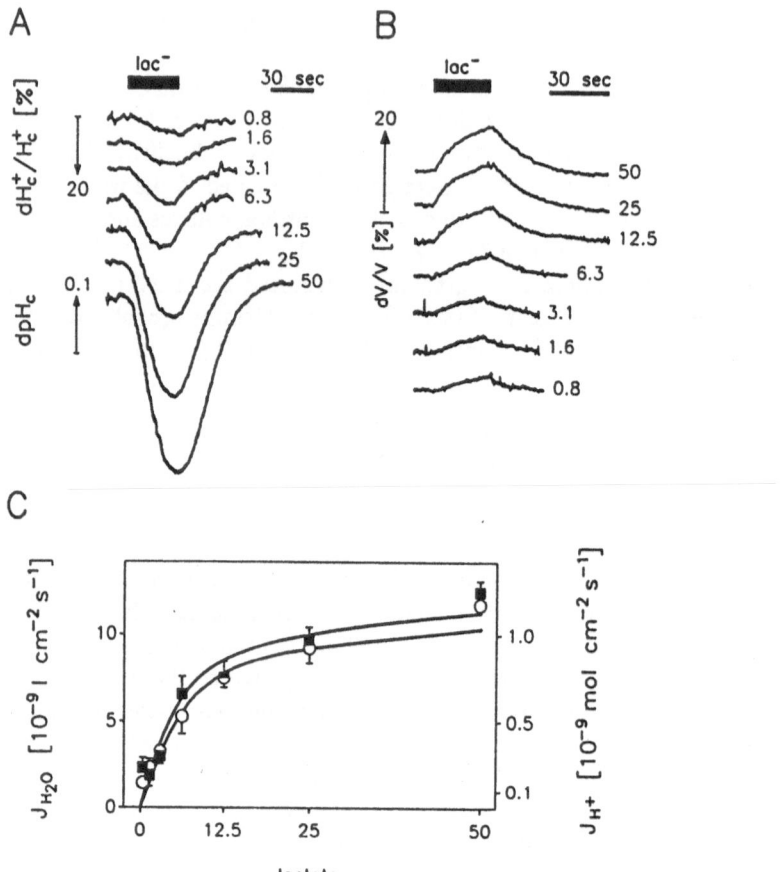

Fig. 9. Coupling ratio and saturation of H$^+$ and water fluxes in the H$^+$/lactate co-transporter. Epithelial cells from the isolated retinal pigment epithelium from Bull-frog were exposed abruptly to test solutions where Cl$^-$ replaced lactate isosmotically, see Fig. 3. This caused immedeate concentration dependent cell acidifications (A) and cell swellings (B). Successive recordings from the same cell. The initial rates of change in cell pH and the buffer capacity of the cell define the influxes $J_H{}^+$. The coupling ratio $J_H{}^+/J_{H2O}$ was relatively constant, around 109 mmol l^{-1} and independent of the lactate concentration (C). Adapted from Zeuthen et al. (1996)

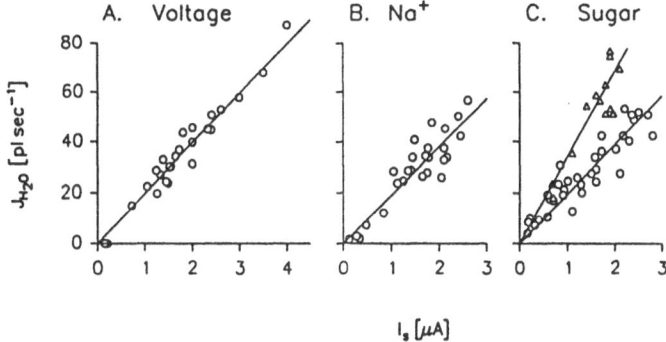

Fig. 10. Coupling ratio of the Na^+/glucose transporter under various conditions. Water transport rates J_{H2O} were obtained at different clamp currents I_s, data from the human Na^+/glucose transporter (hSGLT1) expressed in *Xenopus* oocytes, compare Fig. 4. Steady state values of I_s were obtained at various (A) clamp potentials, (B) external Na^+ concentrations (substitution with choline ions), or (C) external glucose concentrations (substitution with mannitol). In A, the clamp voltage was varied in the range +20 to -100 mV at external Na^+ concentrations of 90 mmol l^{-1} and glucose concentrations of 10 mmol l^{-1}. In B, the Na^+ concentrations were varied in the range 1 to 90 mmol l^{-1} at a clamp potential of -50 mV and glucose concentrations of 10 mmol l^{-1}. In C, the glucose concentration was varied between 0.1 and 20 mmol l^{-1}, at 90 mmol l^{-1} Na^+ and a clamp potential of -50 mV. The average coupling ratio was the same under all three conditions, about 210 water molecules per 2 Na^+ and 1 glucose. The rabbit Na^+/glucose cotransporter (C open triangles) transported more water per 2 Na^+ than the human transporter; in the present example 323 water molecules. Adapted from Meinild et al. (1998a)

4.3
Saturation

Cotransport proteins have finite rates of turnover. Accordingly, any substrate flux should exhibit saturation at increasing driving forces. In the present context, the most interesting experiments are those where saturation of the fluxes was achieved for increasing osmotic gradients. This is a strong indication of water cotransport, saturation would not be expected if water transport was osmotic.

Saturation of the water flux with osmotic gradients is exemplified in Fig. 11 which shows experiments from the H^+/lactate cotransporter (Zeuthen et al 1996). Very similar results were obtained for the K^+/Cl^- cotransporter, see Fig. 4 and 5 in Zeuthen (1994). The water flux is a sum of two components: a saturable water flux maintained by the cotransporter and an unsaturable water flux through the lipid phase and through other membrane proteins, i.e. aquaporins. Saturation of the cotransported water flux was achieved at

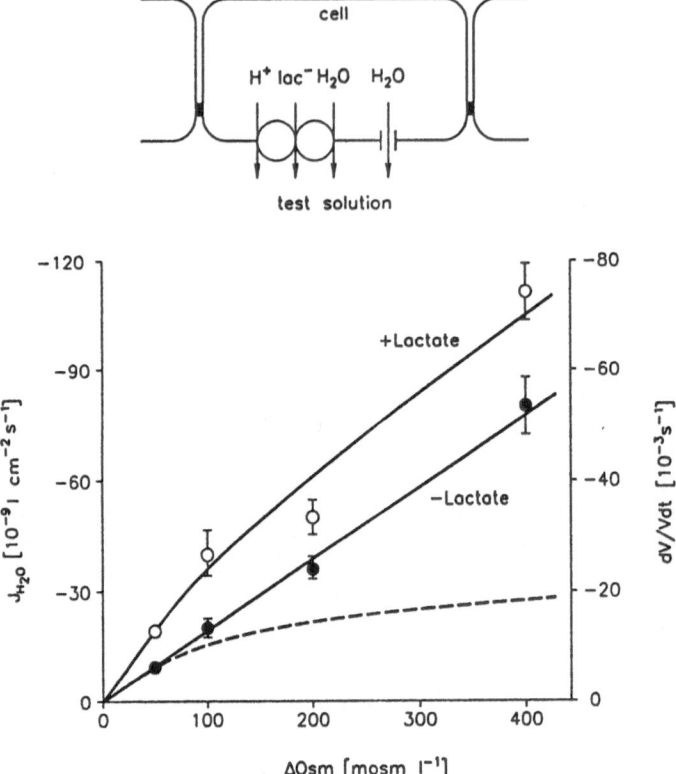

Fig. 11. Saturation of osmotically induced water fluxes in the H$^+$/lactate cotransporter. Open circles represent the water flux J_{H2O} in response to osmotic challenges implemented by additions of mannitol in the presence of lactate, see Fig. 3. Filled circles represent J_{H2O} in the absence of lactate. The difference between the two curves indicates the water transport by the lactate-activated cotransporter (stippled line); it saturates around 200 mosm l^{-1}. In the lactate-free case, water transport across the retinal membrane can be described by a single osmotic water permeability given by the slope of the straight line. Adapted from Zeuthen et al. (1996)

gradients above 200 mOsm l^{-1} for both the H$^+$/lactate and the K$^+$/Cl$^-$ cotransporters. The associated fluxes of the non-aqueous substrates also exhibited saturation around this value. In case of the retinal pigment epithelium, the passive component of the water flux was predominantly through the lipid bilayer; this epithelium contains no known aquaporins (Hamann et al 1998).

   Saturation was also achieved when downhill fluxes of the non-aqueous substrates constituted the driving force. For example, the fluxes of K$^+$, Cl$^-$, and water in the K$^+$/Cl$^-$ cotransporter saturated with increasing K$^+$ gradients, Fig. 8. The H$^+$ and water fluxes in the H$^+$/lactate cotransporter saturated with increasing lactate gradients, Fig. 9. Taken in isolation, however, such findings cannot be taken as evidence for cotransport of water. If the water fluxes were coupled to the fluxes of the non-aqueous substrates by, say, an osmotic mechanism inside the protein, then the water fluxes would saturate when the fluxes of the non-aqueous fluxes saturated, see Section 7.

It is well established that the Na^+/glucose transporter exhibits saturation at increasing gradients of Na^+, sugar and electrical potential (Parent et al. 1992, Wright et al. 1994). The experiments discussed in Section 2.3 were in fact performed with saturating values of glucose and Na^+ concentrations, and of electrical potential. The observation that the coupling ratio of Na^+ fluxes to water fluxes is constant, Fig. 10, shows that the water transport is also saturated under these conditions. Unfortunately, due to the inherent Na^+ and voltage gradients, this cotransporter is poised too far from equilibrium to allow experiments where saturation is achieved by variation in the osmotic gradient.

4.4
Selectivity and Inhibition

In accordance with the cotransport mode, water transport was specifically dependent on the presence of the non-aqueous substrates. The ability of the K^+/Cl^- cotransporter to transport water was dependent on the presence of Cl^-. If Cl^- was replaced by gluconate, thiocyanate, SCN^-, or NO_3^- cotransport of water was abolished. In this case, transport across the apical membrane of the choroid plexus was entirely osmotic, see Fig. 4 in Zeuthen (1991b). Removal of K^+ was not attempted. Furosemide (10^{-4}mol l^{-1}) or Bumetanide (10^{-5} mol l^{-1}) abolished the K^+/Cl^- dependent water transport.

Water transport by the H^+/lactate transporter was dependent on the presence of lactate or other transported monocarboxylates, for example pyrovate (Hamann et al. 1996). In lactate free solutions, all water transport across the retinal membrane of the retinal pigment epithelium was osmotic, Fig. 11. Unfortunately, there is no good specific inhibitor of the H^+/lactate transporter.

Both active and passive (Section 6) water transport by the Na^+/glucose transporter was reversibly inhibited by phlorizin, 100 µmol l^{-1} (Loo et al. 1999b, Meinild et al. 1998a).

5
Thermodynamics of Water Cotransport

In water cotransport, a well-defined number of water molecules participates as a substrate on equal footing with the other substrates. In this case, the relations between the chemical or electrochemical potential differences of the various substrates can be assessed from Gibbs' equation (Zeuthen 1996). The equation applies irrespective of the underlying molecular model.

5.1
$K^+/Cl^-/H_2O$ Cotransport

If water is cotransported by the K^+/Cl^- cotransporter we may talk about $K^+/Cl^-/H_2O$ cotransport. Accordingly, Gibbs' equation predicts equilibrium across the membrane when

$$K^+_c * Cl^-_c * [C_{w,c}]^n = K^+_t * Cl^-_t * [C_{w,t}]^n \qquad (Eq. 1)$$

c indicates the *cis* compartment and t the *trans* compartment, C_w is the concentration of water and n is the number of water molecules transported for each pair of K^+ and Cl^-. C_w is proportional to $\exp(-osm/n_w)$ where osm is the osmolarity in units of osm l^{-1} and n_w is the number of moles of water per liter, about 55. Even if n is as high as 500, the ratio of the respective terms in C_w is only 2.3 for a transmembrane osmotic difference of 100 mosm l^{-1}. With typical intracellular values for K^+ and Cl^- of 80 and 40 mmol l^{-1} in the *cis* compartment and of 2 and 110 mmol l^{-1} in the *trans* compartment, Eq. 1 shows that a transmembrane osmotic difference of 320 mosm l^{-1} can be balanced. It follows that under physiological conditions the ion concentrations are the main determinants of the direction of cotransport. This theoretical prediction agrees with the experimental finding that it takes osmotic differences of 200 to 300 mosm l^{-1} to stop the water transport mediated by the K^+/Cl^- cotransporter in the epithelial cells of choroid plexus, see Fig. 8 in Zeuthen (1994). Interestingly, this value is also close to the upper limit of osmotic gradients that leaky epithelia can transport against, see Section 9.

A related question is: will there be enough energy in the dissipative flux of KCl to drive the uphill flux of water? The dissipated energy can be estimated from the product of the flux of KCl and the difference in chemical potential of the two ions across the membrane, $= J_{KCl}*RT*\ln(K_c*Cl_c/K_t*Cl_t)$. As an example, leaky epithelia from amphibia typically transport isotonic volumes at a rate of 10 µl hour^{-1}cm^{-2} or $2.7*10^{-9}$ l sec^{-1} cm^{-2}. An isotonic flux will represent 110 mmoles of cations per liter of water. Thus, it is reasonable to set the efflux of KCl via the cotransporter (J_{KCl}) to about $0.3*10^{-9}$ mol cm^{-2} sec^{-1}. With intracellular (c) concentrations of K$^+$ and Cl$^-$ of 80 and 40 mmol l^{-1} and extracellular (t) values of 2 and 110 mmol l^{-1}, the dissipated energy amounts to $2*10^{-6}$ joule sec^{-1} cm^{-2}. To move a flux of water of $2.7*10^{-9}$ l^{-1} sec^{-1} cm^{-2} against a gradient of 100 mosm l^{-1} requires an input of $2.7*10^{-9}$ l sec^{-1} cm^{-2} *2.24 atm = $0.6 \cdot 10^{-6}$ joule sec^{-1} cm^{-2}. A comparison between the dissipated and the required energies shows clearly that transepithelial transport of water against gradients of 300 mosm l^{-1} is possible if the dissipated energy is utilized for water transport (Zeuthen 1994).

The measurements outlined in Section 4 can be used to predict how much water is transported by the KCl transporter across the exit membrane of the choroid plexus epithelium in steady states under physiological conditions. The water flux J_{H2O}, will be proportional to the difference in chemical potential between the two sides of the membrane. For conditions close to steady states:

$$J_{H2O} = B * (K^+_c * Cl^-_c * [C_{w,c}]^n - K^+_t * Cl^-_t * [C_{w,t}]^n) \qquad \text{(Eq. 2)}$$

Where B is a constant. Since C_w is proportional to $\exp(-osm/n_w)$ differentiation gives:

$$\Delta J_{H2O} / \Delta osm_t = B * K^+_t * Cl^-_t * n / n_w * \exp(-n * osm_t / n_w) \qquad \text{(Eq. 3)}$$

The value of $\Delta J_{H2O} / \Delta osm_t$ can be obtained experimentally, Fig. 4 in Zeuthen (1994) and amounts to $0.6*10^{-4}$ cm sec^{-1}. All factors in Eq.3 are then known except for B which calculates as $1.8*10^{-2}$ cm sec^{-1} (osm l^{-1})$^{-2}$. With this value of B inserted into Eq. 2 the cotransport component of the water transport can be assessed. With equal intra- and extracellular osmolarities of about 300 mosm l^{-1} and the above concentrations of K$^+$ and Cl$^-$, J_{H2O} is calculated as 3.6 nl cm^{-2} sec^{-1} with the direction out of the cell. This value is within the range of net transport rates found for leaky epithelia from amphibia: 0.9 nl cm^{-2} sec^{-1} in Necturus proximal tubules to 4.1 nl cm^{-2} sec^{-1} in Necturus gall bladder epithelium, for references see Zeuthen (1996) It is very close to that observed in frog choroid plexus when surface foldings are taken into account (Wright et al. 1977). J_{H2O} calculated from Eq.2 as a function of the osmotic difference across the cell membrane is plotted in Fig. 12. The calculated J_{H2O} is rather insensitive to osmotic differences of the order of 30 mosm l^{-1}. It is also seen that the water flow is stopped by an osmotic difference of about 300 mosm l^{-1} in agreement with Eq. 1.

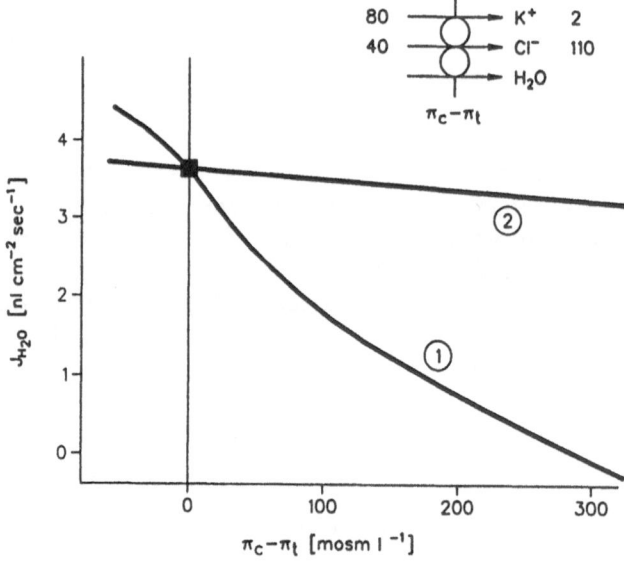

Fig. 12. Quantitative evaluation of water transport by the K^+/Cl^- cotransporter. The $K^+/Cl^-/H_2O$ cotransporter of the *Necturus* choroid plexus epithelium is poised between the cellular (or *cis*) compartment (c) and the *trans* (t) compartment which have K^+ and Cl^- concentrations of 80 and 40 mmol l^{-1} and 2 and 110 mmol l^{-1} respectively (insert). The rate of water transport in steady state, J_{H2O}, is calculated from Eq. 2 as a function of the difference in osmolarity between the *cis* and the *trans* compartment, $\pi_c - \pi_t$. Curve 1 is calculated assuming that the *cis* compartment is made hyperosmolar relative to the *trans* compartment by an impermeable osmolyte. In this case about 300 mmol l^{-1} is required to stop transport. Curve 2 is calculated assuming NaCl is removed from the *trans* compartment. In this case the lowered Cl^- concentration in the *trans* compartment speeds up the rate of cotransport and counteracts the unfavorable effect of the osmotic gradient. Adapted from (Zeuthen 1996)

5.2
H^+/Lactate/H_2O Cotransport

According to Gibbs' equation, equilibrium exists across the retinal membrane of the pigment epithelium (Fig. 3) when

$$H^+_c * lac_c * [C_{w,c}]^n = H^+_r * lac_r * [C_{w,r}]^n \qquad \text{(Eq. 4)}$$

H^+, lac and C_w are the activities of hydrogen ions, lactate ions, and of water; r denote the retinal solution and c the cellular compartment; n is the number of water molecules transported for each pair of H^+ and lactate. As C_w is proportional to $\exp(-osm/n_w)$ the equilibrium condition can be expressed more conveniently as

$$H^+_c * lac_c / H^+_r * lac_r = \exp[\, n / n_w * (osm_c - osm_r)] \qquad \text{(Eq. 5)}$$

where osm are osmolarities in osm l^{-1} and n_w is the number of moles of water per liter (about 55). Consider the physiological situation where pH_r equals 7.4 and pH_c equals 7.2. The lactate concentration in the retinal space under physiological conditions is about 7 mmol l^{-1} (Heath et al. 1975, laCour et al. 1994). The coupling ratio of 109 mmol l^{-1} implies an n of 500 (Zeuthen et al. 1996). If we assume that the intracellular lactate concentration is low, say 1 mmol l^{-1}, it would take a retinal hyperosmolarity (osm_r - osm_c) of 160 mosm l^{-1} to balance the H^+ and the lactate gradients. In other words, the cotransporter would mediate an influx of water as long as the retinal hyperosmolarity was less than 160 mosm l^{-1}. Inspection of Eq.5 shows that the cotransport is very sensitive to the intracellular parameters. For the mechanism to be useful, H^+ and lactate must be removed across the choroidal membrane and into the circulation.

It is possible to predict the rate of water transport by the H^+/lactate cotransporter in a manner analogous to that of the K^+/Cl^- cotransporter, Eq. 2 and 3. The presence of lactate in the retinal space will activate water cotransport and give rise to a flux of $7*10^{-9}$ l cm^{-2} (Zeuthen et al. 1996). This is more than three times the basal rate measured under lactate free conditions in the retinal pigment epithelium (Hughes et al. 1984, Frambach et al. 1985).

5.3
Na$^+$/Glucose/H$_2$O Cotransport

Under physiological conditions or when expressed in *Xenopus* oocytes, the Na$^+$/glucose transporter is poised far from electrochemical equilibrium; the inward flux will be driven by both the Na$^+$ gradient and the electrical potential gradient. For the intestinal Na$^+$/glucose transporter (SGLT1) the driving force is enhanced by the fact that two Na$^+$ are transported for each turnover of the cotransporter (Parent et al. 1992, Chen et al. 1995, Mackenzie et al. 1998). If water is cotransported by the Na$^+$/glucose cotransporter, we may talk about Na$^+$/glucose/H$_2$O cotransport. Gibbs' equation predicts equilibrium when

$$RT\ln(Na_o/Na_c)^2 + RT\ln(G_o/G_c) + RT\ln(C_{w,o}/C_{w,i})^n = 2F(\Psi_c - \Psi_o) \qquad \text{(Eq. 6)}$$

o defines the outside and c the cellular compartment, C_w is the water concentration, n is the coupling ratio, which for the human SGLT1 equals 210, and Ψ the electrical potential (Meinild et al. 1998a). Consider the case where the outside Na^+ concentration is ten times higher than the intracellular, the glucose concentrations (G) the same, and the membrane potential difference -50 mV. It can be calculated that the inward water flux would proceed in spite of adverse osmotic differences of up to 2300 mosm l^{-1}. In case the intracellular concentration of glucose was ten times higher than on the outside, the difference would be 1650 mosm l^{-1}. It follows that the resulting water transport is quite insensitive to the small osmotic differences that may occur physiologically, in agreement with the findings presented in Fig. 5 and 18A.

6
Passive Water Permeability in Cotransport Proteins

In addition to their recognized functions, many membrane proteins act as passive water channels. That is, they have a well defined water permeability, L_p, and mediate a passive water flux given a transmembrane osmotic difference. This dual function is found among the cotransporters, the aquaporins (AQP), the glucose uniporter (GLUT1), and the cystic fibrosis transmembrane conductance regulator (CFTR), references below. A comparison between the properties of water transport among these groups, such as L_p per protein, activation energy (E_a), reflection coefficients (σ), and gating mechanisms illustrates some molecular properties of the passive pathways. It turns out that the putative pore in cotransport proteins is difficult to envisage in models based exclusively on equivalent pore diameters and pore lengths.

The aquaporins (AQP) which are members of the MIP-family (major intrinsic protein) can be divided into two groups. A group which is predominantly selective to water, for example, AQP no. 0, 1, 2, 4, 5 and γTIP (tonoplast intrinsic protein), and a group which is permeable to water and larger substrates such as glycerol, for example AQP 3, 7 and 9. For reviews see (Agre et al. 1998, Crispeels and Maurel 1994). Single channel water permeabilities are comparatively large (in units of 10^{-14} cm^3 sec^{-1}): For AQP0: 0.015 to 0.25 (Zampighi et al. 1995, Yang and Verkman, 1997); AQP1: 1.4 to 6 4 (Zeidel et al. 1992, Zampighi et al. 1995, Yang and Verkman, 1997); AQP 2: 3.3; AQP3: 3.1; AQP4: 24 and for AQP5: 5.0, (Yang and Verkman 1997). For aquaporins permeable to substrates other than water, there is good evidence that the water and substrate molecules share the putative aqueous pore. In AQP3 (Meinild et al. 1998b, Zeuthen and Klaerke 1999), AQP7 (Ishibashi et al. 1997, Kuriyama et al. 1997), and in AQP9 (Tsukaguchi et al. 1998) the reflection coefficients (σ) for a wide range of small hydrophilic molecules have been found to be much lower than one. Water permeation is characterized by low activation energies, 3–4 kcal mol^{-1} (Preston et al. 1992), 6–7 kcal mol^{-1} (Meinild et al. 1998b). In analogy, the activation energy for glycerol transport is also low (Ma et al. 1994, Maurel et al. 1994, Zeuthen and Klaerke 1999). These findings have led to a model in which water and other substrates are thought to permeate aquaporins by forming a succession of hydrogen bonds (Zeuthen and Klaerke 1999).

The glucose uniporter GLUT1 is permeable to water (Fishbarg et al. 1990) with a single channel water permeability of about 6 to 60% that of AQP1. Experiments in the human erythrocytes indicate that the activation energy is higher than expected from aqueous diffusion (Zeidel et al. 1992).

When CFTR is expressed in *Xenopus* oocytes and activated via cAMP the oocyte membrane exhibits an increased permeability for water (Hasegawa et al. 1992). Resent investigations suggest that the water permeation and the Cl$^-$ conductance take place at different locations of the protein (Schreiber et al. 1997).

The passive water permeability of cotransport proteins has been studied in details in the intestinal Na$^+$/glucose transporter (Zampighi et al. 1995, Loo et al. 1996, Loike et al. 1996, Zeuthen et al. 1997, Meinild et al. 1998a, Loo et al. 1999b), in the Na$^+$/Cl$^-$/GABA cotransporter (Loo et al. 1999b), and in the Na$^+$/dicarboxylate and the H$^+$/amino acid transporter (A.-K. Meinild, unpublished). The Na$^+$/glucose transporter is characterized by a single channel water permeability which is about 100 times lower than that of AQP1 (Zampighi et al. 1995, Loo et al. 1999b), and by a low activation energy, 5 kcal mol^{-1} (Loo et al. 1996, Loo et al. 1999b). Yet, they are permeable to small hydrophilic molecules; the reflection coefficient for formamide is about 0.3 in the human Na$^+$/glucose cotransporter (Meinild et al. 1998a). In comparison, the AQP1 is mainly selective to water (Meinild et al. 1998b), it has a minute permeability to glycerol but it is impermeable to formamide.

Formalisms based on equivalent pore radius and pore length would appear to be inadequate to reconcile the observations in AQP1 and the Na$^+$/glucose transporter. In such models the AQP1 is as narrow as it can be, it only lets through the smallest molecule, water. The Na$^+$/glucose cotransporter which has a single channel water permeability which is 100 times smaller would need an aqueous pore which was about 100 times longer than the AQP1. This seems unlikely. It is also unlikely that the water permeability of the Na$^+$/glucose transporter is determined by stochastic openings and closures (gating). Such a mechanism should reveal itself by having high activation energies.

An alternative strategy would be to interpret the data by means of an Eyring energy barrier model (Glasstone et al. 1941, Hille 1992). On this model, the water molecule permeates by a series of jumps, the energy barriers of which are determined by the chemical bonds between the water molecule and specific sites in the pathway, for example hydrogen bonds (Zeuthen and Klaerke 1999). A water molecule transversing the pore in the Na^+/glucose cotransporter might encounter more and stronger bonds than a water molecule transversing the AQP1.

In addition to the passive water permeability, the Na^+/glucose transporters have, in the absence of sugars, a passive leak current carried by cations, Na^+ or H^+ (Wright et al. 1996, 1998). This leak current is determined by a different mechanism than that of the passive water flux. Or more precisely, it depends on a different set of conformational changes in the protein than does the passive water permeability (Loo et al. 1999b). The leak current is saturable, depends on the type of cation, and has high activation energies, E_a, 28 kcal mol^{-1}. In comparison, the water permeability does not saturate and is independent of the type of cation.

7
Mechanisms of Molecular Water Pumps

How could the structural features and conformational changes in cotransport proteins give rise to secondary active water transport or even cotransport of water? Two distinctly different types of models have been suggested, the mobile barrier model (Zeuthen 1994, Zeuthen and Stein 1994) and the osmotic coupling model (Zeuthen 1991b and Zeuthen and Stein 1994). Before discussing the models, it will be useful first to have a look at the general structure of some cotransport and channel proteins, and to review what is known about the nature and amounts of water attached to enzymes.

There are evidences that the basic structural feature of proteins engaged in transmembrane transport is a number of membrane-spanning α-helices. For cotransporters the number lies around 12, for references see (Zeuthen 1996), for the Na^+/glucose cotransporter 14 (Turk et al. 1996) and 15 has been suggested (Eskandari et al. 1998). In many models the α-helices form a pore through which substrate pass from one side of the membrane to the other. The diameter of this pore varies being wide for most of its length but with a narrow short portion that contains binding sites for the dehydrated substrates and thus acts as a barrier. The hydrophilic loops that connect the α-helices may fold back into the pore and be constituents of this barrier. This scheme applies broadly for bacterial porins (Nikido and Saier 1992), ion channels (Latorre and Miller 1983, Doyle et al. 1998), water channels

(Agre et al. 1998), and cotransporters, for references see (Zeuthen 1996, Wright et al. 1996). The model is in agreement with the one suggested by Mitchell (1957).

Conformational changes in enzymes involve movements of complete protein domains relative to each other. This changes the size of the proteinaceous surfaces covered with loosely bound water and thereby the amount of water held by the protein. Conformational changes may also change the size of water filled cavities in the proteins. The effects can be studied in experiments where enzymatic activity is monitored as a function of the chemical potential for water in the external medium (Parsegian et al. 1993). It has been applied to a number of enzymes, both membrane-bound and water-soluble, and has revealed a wide range of numbers for exchangeable water molecules, ten to more than a thousand, Table 2. Such numbers are not inconceivable neither when the water molecules are enclosed in cavities nor if they are attached to the protein as surface water. 500 molecules of bulk water occupy a volume of 15300 A^3. With a molecular weight of the K^+/Cl^- cotransport proteins of 260 kD i.e. (Cherksey and Zeuthen 1987), this constitutes only about 3% of the protein volume. In studies of hemoglobin (Colombo et al. 1992), one surface-bound water molecule was found to occupy 10 A^2. In this way, 500 water molecules would occupy a surface of 5000 A^2, equivalent

Table 2. Number of water molecules (n) exchanged with the external solution per cycle of enzymatic action

Protein	Process	n	Reference
Voltage-gated anion channel	Opening/closing	660–1320	Zimmerberg and Parsegian 1986
KCl -cotransporter	H_2O cotransport	500	Zeuthen 1994
hSGLT1	H_2O cotransport	210	Meinild et al. 1998a
Hexokinase	Glucose binding	100	Steitz et al. 1981, Rand and Fuller 1992
α-Amylase	Glucose binding	90	Qian et al. 1995
Hemoglobin	O_2 binding	60–75	Bulone et al. 1991, Colombo et al. 1992
K^+ channel	Opening/closing	40–50	Zimmerberg et al. 1990
Cytochrome C oxidase	Electron transfer	10	Kornblatt and Hoa 1990, Kornblatt et al. 1998, Kornblatt 1998

to the surface of a 100A long, 15A wide circular pore. Such considerations are not incompatible with direct crystallographic evidence. Studies of the sugar binding enzyme α-amylase have shown that between 5 and 10% of the protein volume is occupied by water. This is equivalent to 300 to 500 water molecules; about 90 of which are exchanged with bulk water during conformational changes (Quin et al. 1995).

7.1
Mobile Barrier Model

The mobile barrier model is based upon several well established phenomena observed in aqueous enzymes. In the model, these phenomena take place in a specific order so as to give rise to vectorial transport of water and the other substrates, Fig. 13A. The non-aqueous substrates gain access to the binding sites by migration in an aqueous cavity which opens to the *cis* side of the protein while a barrier separates the substrates and the *trans* compartment. Migration of K^+ in an aqueous funnel has recently been established for a K^+ channel. It was argued that the water served to shield the charges of the ion prior to binding (Doyle et al. 1998). Substrate binding results in a conformational change that causes the barrier to open exposing the bound substrates to the *trans* compartment. Simultaneously, another barrier closes in order to separate the substrates from the *cis* compartment. The substrates are then released into the *trans* compartment. In analogy to the aqueous enzymes the cavity finally closes and its water delivered into the *trans* compartment. From this closed configuration the protein opens again to the *cis* side. In other words, it takes up water from the *cis* side and is ready for a new transport cycle. The water released at the closure of the cavity can be the bulk type of water or it can be loosely bound surface water lining the cavity.

On this model, the transport of the non-aqueous substrates cannot take place without the water. In other words, water is also a substrate. This means that the osmotic work in transporting the volume of water contributes to the overall energy balance as does the chemical and electrical work (Eq. 1).

The properties of the mobile barrier model as a molecular water pump was alluded to by P. Mitchell (1990): "If as seems likely, the mobile barrier mechanism involves the opening and closing of a cleft on either side of a substrate-binding domain, one might expect considerable hydrodynamic action as aqueous medium was sucked in and squirted out by the crevices".

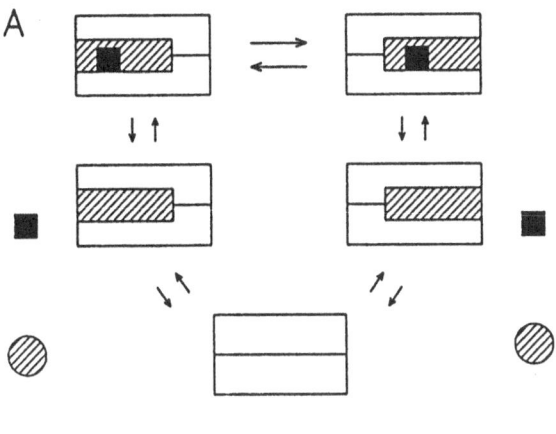

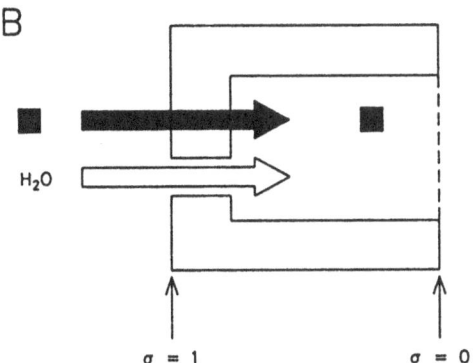

Fig. 13. Molecular models for molecular water pumps. **A.** The mobile barrier model (Mitchell 1957) adapted for cotransport of non-aqueous substrates (■) and water (hatched) in cotransport proteins. Hydration of the access volume allows binding of the non-aqueous substrates. This induces conformational changes that shift the permeability barrier from one end of the protein to the other; subsequently the non-aqueous substrate is delivered to the *trans* side. For vectorial transport of water to ensue, it is assumed that the protein also exists in a closed state where the access volumes are relatively small. Adapted from (Zeuthen 1994). **B.** The osmotic coupling model. The non-aqueous substrate (■) is transported from the *cis* compartment into a cavity in the protein. This causes the cavity to become hyperosmolar relative to the outer solutions. The cavity is limited to the *cis* side by a semi-permeable membrane, reflection coefficient $\sigma = 1$. The *cis* barrier has a passive water permeability and water is transported by osmosis from the *cis* side into the cavity. The *trans* side of the protein is characterized by a low reflection coefficient ($\sigma \cong 0$) and by a certain permeability to the substrate. The model is based on the two barrier model (Curran and Macintosh 1962, Patlak et al. 1963), see also Zeuthen and Stein (1994)

7.2
Osmotic Coupling Model

In the osmotic coupling model an internal cavity of the protein functions as a coupling compartment in the sense of the two barrier (or three compartment) model, Fig. 13B (Curran and Macintosh 1962, Patlak et al. 1963). The first barrier separates the *cis* compartment and the cavity. This barrier constitutes the selectivity filter for the non-aqueous substrates. It has an osmotic water permeability, L_p, and acts as a semipermeable membrane with a reflection coefficient (σ) close to one. The other barrier is located at the *trans* side of the protein and discriminates only little between the non-aqueous substrates and water. Consequently, it has a reflection coefficient close to zero. In addition, it must exert a certain restriction on the efflux of the non-aqueous substrates. Transport of the non-aqueous substrates into the cavity increases the osmotic concentration, giving rise to an osmotic flow of water across the first barrier. The resulting increase in hydrostatic pressure in the cavity causes a hydrostatically driven flow across the second barrier. Theoretical analysis of this type of models (Patlak et al. 1963, House 1974) show that a wide range of transport tonicities can be achieved with different choices of water permeabilities and reflection coefficients. Certain choices of parameters lead to hydrostatic pressures in the middle compartment that are too high to be sustained by compartments consisting of cell membranes. This would be a minor problem for the model when applied to proteinacious structures. The walls of the cavity in the protein is held together by molecular bonds, which are expected to be able to sustain large pressures.

There are two inherent problems with the model when applied to proteins. Obviously, the concentrations of the non-aqueous substrates in the cavity cannot rise above a value where the (electro)chemical potential difference across the *cis* barrier is zero, in this case the flow would simply stop. This puts an upper limit to the osmolarity in the cavity. Second, the *trans* barrier of the protein must be able to restrict the escape of the non-aqueous substrates in order to maintain a high osmotic concentration in the cavity. This must be achieved without increasing the reflection coefficient of this barrier.

7.3
Comparison Between Models and Experiments

So far, the mobile barrier model explains more data than the osmotic coupling model. In the mobile barrier model the ratio of any two fluxes is an inherent property of the cotransport protein. This gives a simple explana-tion of why the measured coupling ratios are independent of the external parameters or the driving forces, see Section 4.

The strength of the osmotic coupling model lies in its explanation of the *raison d'etre* for the passive water permeability of the cotransport protein (Section 6). On the other hand, the constancy of the coupling ratio is difficult to explain in this model. The concentration in the middle compartment is a complicated function of the external concentrations and the rates of trans-port (Patlak et al. 1963). Accordingly, it would be necessary to assume that the transport parameters of the *cis* and *trans* barriers of the protein, i.e. the water permeabilities and the reflection coefficients, were complicated func-tions of the external parameters in order to explain the constancy of the coupling ratios. It would be particularly difficult to see how the osmotic coupling model could explain experiments where an imposed osmotic gra-dient causes fluxes of the non-aqueous substrates, i.e. Fig. 7. A flux of water through the protein could hardly lead to sufficient accumulations of non-aqueous substrates on one side of the *cis* barrier or depletion on the other side to reach the required driving forces.

The final choice between the two models might come from measure-ments under equilibrium conditions. Water is a co-substrate in the mobile barrier model. In this case, the reversal potential of the cotransporter would depend on the water chemical potential difference across the membrane. The reversal potential in the osmotic coupling model would be independent of the osmotic difference.

8
Cellular water Homeostasis:
A Balance Between Pumps and Leaks

Cellular water homeostasis is usually considered a result of passive water fluxes. With the introduction of molecular water pumps, cellular water ho-meostasis becomes a balance between molecular water pumps and molecular water leaks, Fig. 14. This gives cells the advantage that they can maintain a relatively constant internal milieu i.e. in order to ensure optimum enzymatic activity despite fluctuations in the tonicity of the environment.

Cellular water homeostasis is particularly interesting in cells from water transporting epithelia. Epithelial cells from gall-bladder, proximal tubule, and choroid plexus survive and retain their functions in face of large dilu-tions of the external medium (Hill and Hill 1978, Whittembury and Hill 1982, Zeuthen 1982, 1983). Amphibian gall-bladder, for example, transports

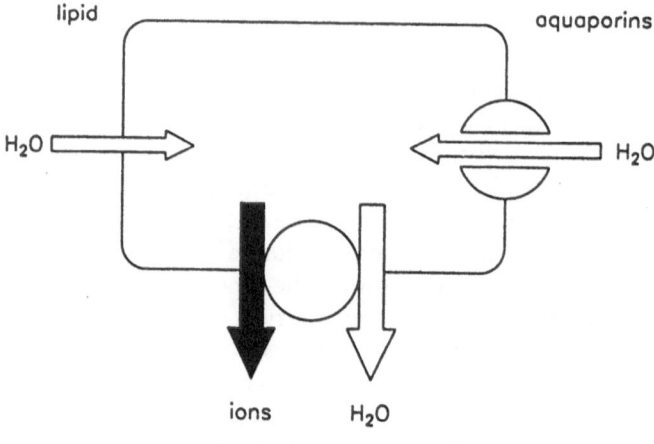

Fig. 14. Cellular water homeostasis as a balance between molecular water pumps and the leaks formed by aquaporins and the lipid bilayer

for hours in solutions of osmolarities of 40 mOsm l^{-1}, a five times dilution compared to normal. Interestingly, the rate of volume transport is five times faster than normal under these conditions (Zeuthen 1982). There appear to be no specific ultrastuctural changes apart from a pronounced distention of the lateral intercellular spaces caused by the increased rates of volume transport (Bundgaard and Zeuthen 1982). It should be emphasized that under dilute, as well as under normal conditions the epithelial cells are surrounded by the same osmolarity on all sides. The secreted solution is isotonic to the bathing solution and, most importantly, the solution in the lateral intercellular spaces has the same osmolarity as the secretion (Zeuthen 1983, Ikonomov 1985). A central question therefore: Will the cell cytoplasm remain isotonic to the external solution in case of external dilutions? If the answer is yes, the dilution of the intracellular milieu would compromise the function of the intracellular enzymes (Lubin 1982, Zimmerman and Harrison 1987, Zimmerman and Minton1993). If, on the other hand, the cell retained its normal osmolarity despite the external dilution, the cell membrane must be able to maintain a water chemical potential difference. In that case, we would have to give up the dogma that cells are in osmotic equilibrium with the surroundings (Steinbach 1962, Zeuthen 1996). Experiments with ionselective microelectrodes and micro-osmometry show that the epithelial cells conform to the latter alternative: When the cells are transporting in dilute external solutions they are hyperosmolar relative to the surrounding solutions (Zeuthen 1981, 1982, 1983). For example, in cells from *Necturus* gall-bladder adapted to transport in solutions of 50 mOsm l^{-1} the

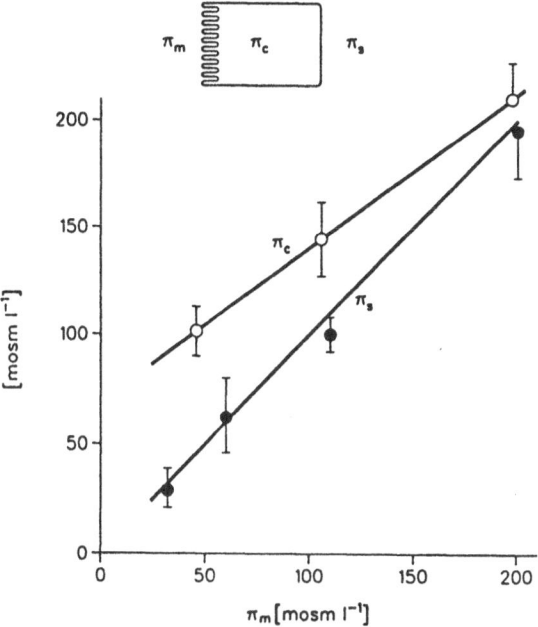

Fig. 15. Relation between intracellular and extracellular osmolarity in gall-bladder cells. *Necturus* gall-bladders were adapted to steady-state transport with different solutions on their mucosal side, osmolarity π_m. The secreted solution had the same osmolarity, π_s, as the solution in the lateral intercellular spaces, and was equal to that of the mucosal bathing solution π_m. Consequently, the cell is surrounded by the same osmolarity, $\pi_m = \pi_s$. The intracellular osmolarity π_c was calculated as the sum of the intracellular concentrations of free K^+, Cl^-, and Na^+ measured with ion selective electrodes. This must be considered a lower estimate of the total intracellular osmolarity. The data were confirmed by micro-osmometry. The cellular hyperosmolarity was abolished by ouabain and lack of oxygen. Adapted from Zeuthen (1981, 1982, and 1983)

sum of the intracellular activities of Na^+, K^+, and Cl^- alone, were equivalent to an intracellular osmolaritiy of about 100 mOsm l^{-1}, see Fig. 15. The intracellular hyperosmolarity required an intact metabolism and a functional Na^+/K^+ATPase. Oxygen removal and ouabain abolished the cellular hyperosmolarity.

The question of whether cells are isosmolar with their surroundings has been discussed intensively over the years. In the mid 1950s it was established that the osmolarities of whole organs such as hearth, muscle, brain and liver were isosmotic to plasma within 1.4%. Measurements on whole kidney were more uncertain due to the hyperosmolarities of the medulla (Conway and Geoghegan 1955, Conway et al. 1955, Appelboom et al. 1958, Maffly and Leaf 1959). The studies showed that what was formerly thought of as cell hyperosmolarities (as much as several 100 mOsm l^{-1}) was a result of artificial tissue autolysis. Unfortunately, the results have been extrapolated to mean that all cells under all conditions are exactly isosmotic with the surroundings. Evidently, it does not hold for osmotic differences of the order of 50 mosm l^{-1} in epithelial cells from leaky epithelia.

The emerging picture is one where water homeostasis arises as a balance between molecular water pumps and molecular water leaks. Take the role of K^+ as an example. The K^+ gradient maintained by the Na^+/K^+ATPase is not simply dissipated. The energy available from the downhill efflux of K^+ is harnessed by the $K^+/Cl^-/H_2O$ cotransporter for the uphill efflux of water, Fig. 14. On this model water enters the cell passively via the lipid bilayer and water channels, i.e. aquaporins.

9
Molecular Water Pumps and Transport Across Leaky Epithelia

Cotransport of water, such as $Na^+/glucose/H_2O$ and the $K^+/Cl^-/H_2O$, could be important building blocks in a general molecular model for vectorial salt-

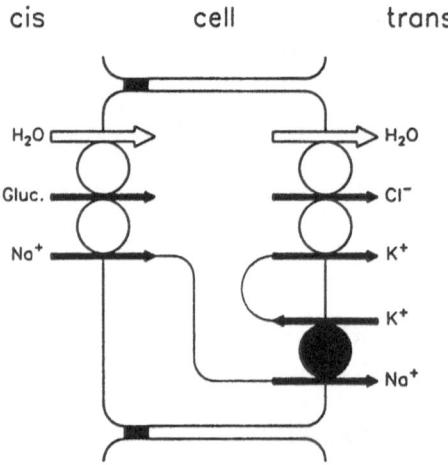

Fig. 16. Coupling between water transport mediated by the $Na^+/glucose$ and K^+/Cl^- cotransporters and metabolism. The Na^+/K^+ ATPase (filled circle) connects the expenditure of metabolic energy with water transport: The influx of water depends on the low intracellular Na^+ concentration and the negative electrical potential; the efflux of water depends on the combined intracellular concentrations of the K^+ and Cl^- ions. A complete molecular model would include additional transporters, i.e. ion and water channels, as well as signal systems for correlation of the various transporters. The two transporters are located to different aspects of the cell. In an epithelium this would give rise to transepithelial water transport, from *cis* to *trans*

water coupling in an epithelial cell, Fig. 16. When combined with a Na^+/K^+-ATPase, the resulting low intracellular Na^+ concentration and negative electrical potential would drive an influx of Na^+ and water, while the combined intracellular K^+ and Cl^- concentrations would energize an efflux of water.

A complete cellular model should include passive water permeabilities, in particular aquaporins, as well as other cotransporters and ion channels. In recent experiments on the defective proximal tubular fluid reabsorbtion in transgenic aquaporin-1 null mice (Schermann et al. 1998), the transepithelial osmotic water permeability was found to be only 20% of that found in wild-type mice while the spontaneous fluid absorbtion rate was only reduced by a factor of two. This implicates that in the proximal tubule about half the water reabsorbtion is mediated via passive transport via aquaporin-1 driven by a transepithelial osmotic difference. But equally important, the result raises the question of how the remaining 50% of the water transport is maintained. It has been suggested that water cotransport could be implicated (Loo et al. 1999a), see also Spring (1999).

The separate roles of aquaporins and molecular water pumps become apparent if the properties of the small intestine are juxtaposed to those of the proximal tubules, Fig. 17. The small intestine has a comparatively low water permeability and any significant presence of aquaporins has not been demonstrated. This absence of aquaporins would tend to give a lower rate of water transport but most importantly for intestinal function it also provides the capacity for uphill transport over a wide range of luminal osmolarities. How this could result from the function of molecular water pumps is discussed below. The proximal tubule has high passive water permeabilities as a result of the presence of aquaporins, in particular aquaporin-1. In this way small transepithelial osmotic gradients can be utilized in order to increase transport rates. The concurrent reduction in the capacity for uphill water transport is not important in view of the well controlled environment of this epithelium.

9.1
Water Transport Across Apical Membranes

Na^+ driven uptake of sugars and amino acids are obligatory processes at the brush border of the small intestine and kidney proximal tubule. The Na^+/glucose transporter from human small intestine was found to transport 210 water molecules for each 2 Na^+ and 1 glucose, Table 1. In humans on a normal diet of, say, 1 mole of glucose a day, this means that about 5 liters of water per day could be reabsorbed by cotransport across the brush border,

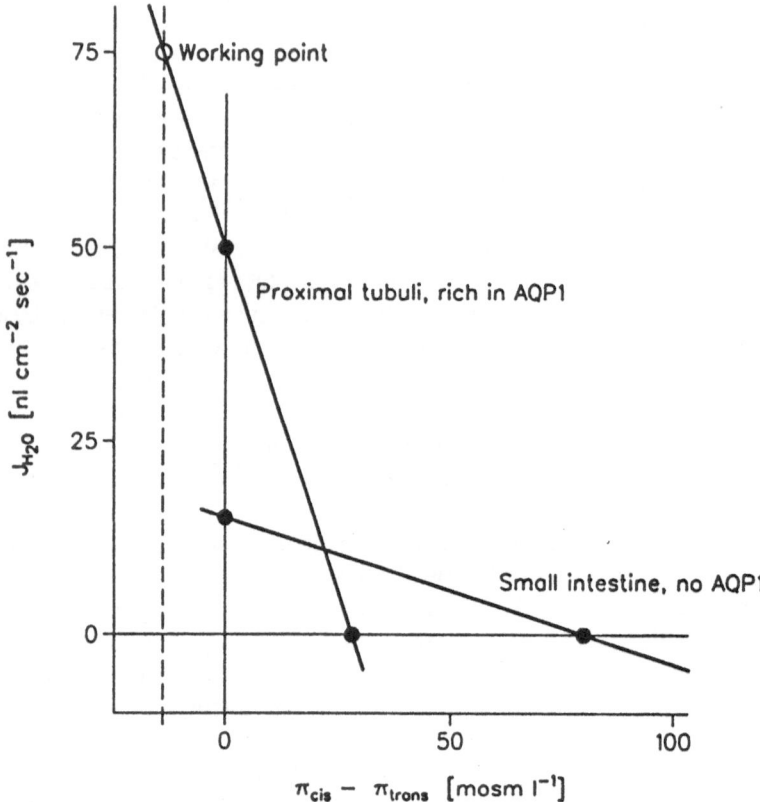

Fig. 17. Transepithelial watertransport, J_{H2O}, as a function of transepithelial osmotic difference $\pi_{cis} - \pi_{trans}$ in the kidney proximal tubule and the small intestine. The proximal tubule is rich in aquaporin 1 and has a high passive water permeability. Consequently, the relation between J_{H2O} and osmotic difference is characterized by a steep slope (high water permeability). The proximal tubule takes advantage of this in order to increase the rate of water transport by creating a solution in the *trans* compartment which is slightly hyperosmolar relative to the tubular (*cis*) solution. In other words, its working point is given by $\pi_{trans} > \pi_{cis}$. It follows that the proximal tubule is unable to transport against larger transepithelial osmotic differences. The small intestine has few, if any, aquaporins. It has therefore a low passive water permeability (shallow slope). This enables it to transport against large osmotic differences. Curves based on data from Green et al. (1991) and Parsons and Wingate (1961), see also Zeuthen (1992), Zeuthen and Stein (1994)

equivalent to about 50% of the total daily uptake (Loo et al. 1996). The small intestine is known to transport a solution isotonic to plasma, in mammals about 300 mosm l^{-1}, or 185 water molecules per solute molecule. The coupling ratio of the Na^+/glucose cotransporter is equivalent to about 70 water

molecules per solute molecule only. This would result in a hypertonic transportate. Accordingly, there must be an additional flux of water. One possibility is that this additional flux is osmotic, mediated by the passive water
permeability of the Na^+/glucose transporter itself (Section 6). The driving
force could arise from the actively transported hypertonic solution which
would increase the osmolarity of the intracellular compartment (Loo et al.
1999b). The measured parameters do not exclude this possibility. Inspection
of Fig. 18A shows that an osmotic difference of about 15 mosm l^{-1} across a
membrane containing Na^+/glucose cotransporters would result in a passive
water flux equal to the cotransported water flux. In this case the total water
flux would equal the one seen in vivo (about 10 liters a day) and the composition would be isotonic.

In a final model of water transport across brush border membranes,
contributions from all types of cotransporters must be considered (Table 1).

9.2
Water Transport Across Basolateral Membranes

The K^+/Cl^- cotransporter appears to be ubiquitous for basolateral type
membranes in epithelial cells: Kidney proximal tubule (Shindo and Spring
1981, Baum and Berry 1984, Guggino 1986, Eveloff and Warnock 1987,
Sasaki 1988), cortical thick ascending limb of Henle's loop (Greger and
Schlatter 1983), gall-bladder (Reuss 1983, Larsson and Spring 1984), and
small intestine (Halm et al. 1985). It is also present in the apical membrane
of the choroid plexus (Zeuthen 1994) in accordance with the fact that this
epithelium is a backward epithelium, where the apical membrane serves as
the exit membrane.

The K^+/Cl^- cotransporter was first described in the red cell (Kregonow
1971) where it has been studied extensively, for a review see (Lauf et al.
1992). The cotransporter has been documented in a variety of other cell
types: endothelial cells (Perry and O'Neill 1993), hepatocytes (Bianchini et
al. 1988), and ascites tumor cells (Aull 1981, Kramhøft et al 1986). There is
general agreement that the transport is electroneutral, independent of Na^+,
and inhibited by loop diuretics. Cell swelling activates K^+/Cl^- cotransport but
the signal pathway is subject to discussion, for a review see Zeuthen (1996).

A central question in regard to water transport across leaky epithelia is
how water can enter the cell across the apical membrane, at least in part, by
osmosis while leaving the cell again across the basolateral membrane. If the
cell is hyperosmolar, the efflux of water must be uphill. In the molecular

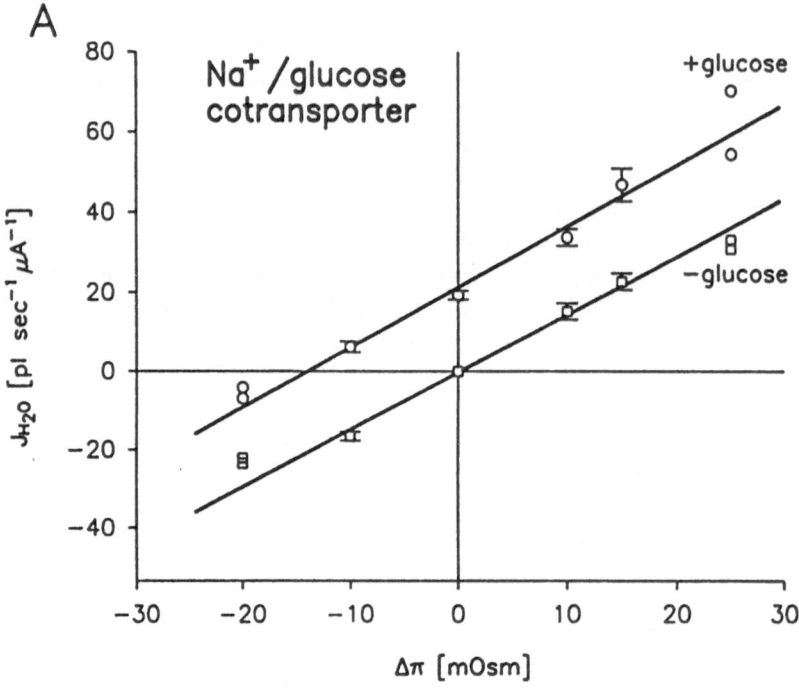

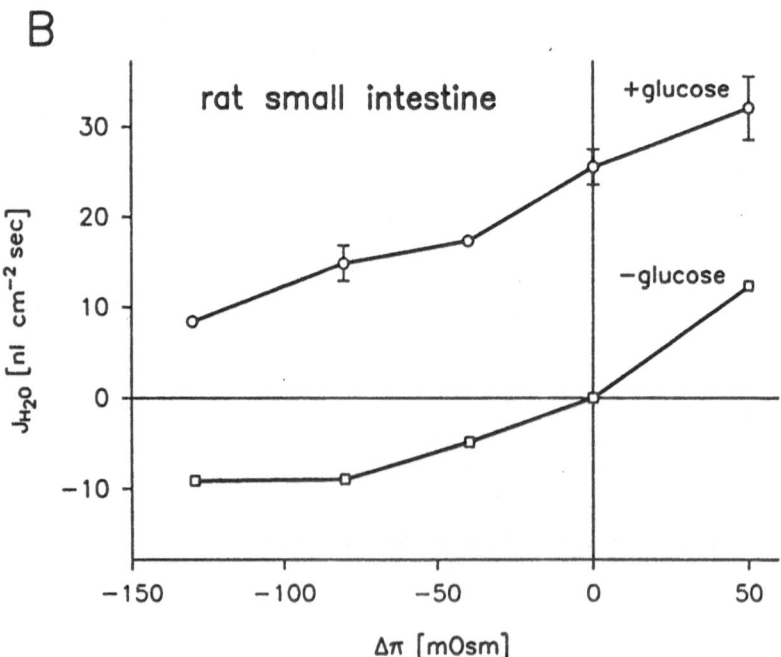

model for trans-epithelial water transport the role of the K^+/Cl^- cotransporter would be to energize the uphill exit of water from the cell. From the data reviewed in Section 5.1, it is clear that the cotransporter is able to maintain sufficient rates of water efflux, see also below.

9.3
Comparison with Data from Whole Epithelia

An epithelial model which incorporates molecular water pumps gives good quantitative predictions of several tissue parameters: The rate of water transport, the ability for uphill transport of water, the effects of sugars, and the reflection coefficient for NaCl (σ_{NaCl}) of the epithelium. Isotonic transport, is not predicted by the model. To achieve this, the dependence of the transporters on external and cellular parameters, i.e. cell swelling and shrinkage, should be included, for a discussion see Zeuthen (1996).

The predictive value of the molecular model in terms of the effects of glucose transport is illustrated in Fig. 18. The presence of glucose provides the small intestine with the ability to transport water in spite of no or even against an adverse transepithelial driving force. The same effect is observed for the isolated Na^+/glucose transporter expressed in *Xenopus* oocytes.

The prediction of rates of water transport, and ability for uphill transport, and reflection coefficients can be illustrated ba a simple model of the chor-

---

Fig. 18. Comparison between water fluxes in the Na^+/glucose cotransporter expressed in *Xenopus* oocytes and water fluxes across intact rat small intestine. A. Water fluxes ($\cong$ volume fluxes) in the human Na^+/glucose transporter from human small intestine expressed in *Xenopus* oocytes as a function of the transmembrane osmotic difference $\Delta\pi$. Data obtained as described in Figs. 4 and 5. The water fluxes were normalized relative to the glucose induced current obtained under saturating conditions. The glucose induced water flux can be divided into a constant contribution of 20.7 pl sec^{-1} (per 1 μA) which is independent of the osmolarity and an osmotic contribution given by an L_p of about $3.6*10^{-6}$ cm $sec^{-1}(osm\ l^{-1})^{-1}$. In case of no sugar present (-glucose), the water flux was entirely osmotic, given by an L_p of $3.4*10^{-6}$ cm $sec^{-1}(osm\ l^{-1})^{-1}$. Accordingly, the glucose induced water flux gives a constant contribution to the total water flux, as evidenced by the constant vertical displacement of the two lines. This is in agreement with the prediction of Gibbs' equation (Eq. 6). The vertical displacement of the two curves gives the coupling ratio of the cotransporter in picoliters of water per μCoulomb. Data from Meinild et al. (1998a). B. Data obtained from intact rat jejunum (Parsons and Wingate 1961). In analogy with the data shown in A, the intestine is characterised by a passive water permeability under glucose free conditions and a constant osmolarity-independent component which depends on the presence of glucose. $\Delta\pi$ is here the transepithelial difference in osmolarity, serosal minus mucosal

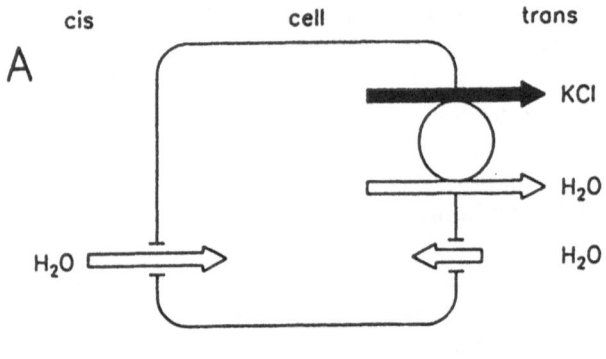

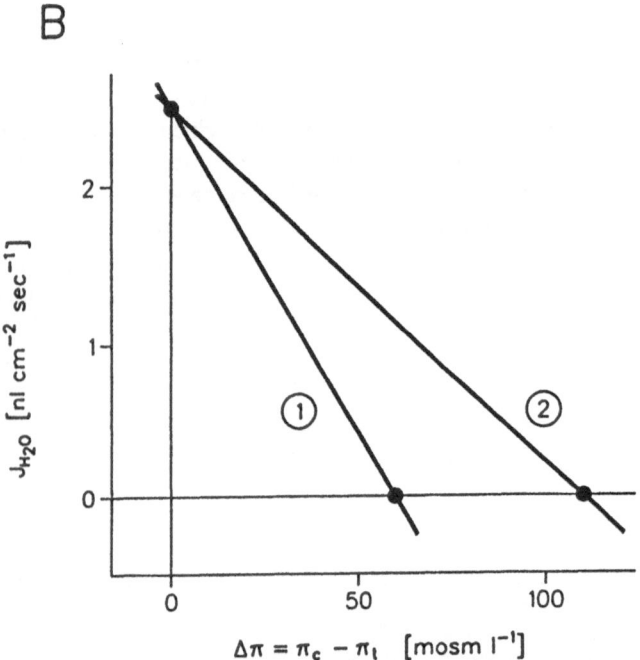

Fig. 19. Quantitative role of $K^+/Cl^-/H_2O$ cotransport in epithelial water transport. A. In this simplified molecular model for epithelial water transport, water is pumped out across the *trans* membrane of the cell by $K^+/Cl^-/H_2O$ cotransport. For simplicity the entry of water across the *cis* membrane is assumed to take place by osmosis. There is a minor passive water permeability in the *trans* membrane. The lengths of the arrows are quantitatively correct for the choroid plexus epithelium from *Nectu-rus*. B. Predictions of the simple molecular epithelial model. Transepithelial water transport J_{H2O} is calculated as a function of the transepithelial osmotic difference implemented by adding mannitol to the *cis* side (curve 1) or by removal of NaCl from the *trans* bath (curve 2). Data from Zeuthen (1994, 1996)

oid plexus epithelium from *Necturus*, where most relevant transport parameters have been determined (Zeuthen 1991a, 1991b and 1994), Fig. 19. For the present purpose it will suffice to assume that the entry of water is osmotic. The water permeability of the entry membrane is has been estimated to be about 10^{-4} cm sec^{-1} (osm l^{-1})$^{-1}$ and the exit of water by the K$^+$/Cl$^-$/H$_2$O cotransporter is given by Eq. 2, Section 5. The exit membrane also has a small passive leak for water given by a water permeability of $0.5*10^{-4}$ cm sec^{-1} (osm l^{-1})$^{-1}$. This leak will cause a minor component of recirculation at this membrane. With these numerical values and the values of K$^+$ and Cl$^-$ concentrations in the cell (100 and 40 mmol l^{-1}) and external fluids (2 and 110 mmol l^{-1}), the calculated rate of transepithelial water transport is 2.4 nl cm^{-2} sec^{-1}, similar to that determined in Bullfrog choroid plexus, 2.8 nl cm^{-2} sec^{-1} (Wright et al. 1977).

The ability for uphill transport in the model cell can be tested by adding mannitol to the *cis* side of the epithelium. In this case the total transepithelial water transport is zero at a transepithelial osmotic difference ($\Delta\pi_{max}$) of 65 mosm l^{-1} (curve 1 in Fig. 19B). Alternatively, uphill transport could be tested by adding NaCl to the *cis* side or diluting the NaCl solution on the *trans* side of the epithelium. This would have the additional effects of establishing a Cl$^-$ concentration difference across the epithelium which would increase the driving force of the K$^+$/Cl$^-$ cotransporter. In this case the transepithelial water flux would be zero when the transepithelial osmotic difference is 110 mosm l^{-1} (curve 2 in Fig. 19B). This prediction of $\Delta\pi_{max}$ is in good agreement with the observation that the choroid plexus epithelium can transport against osmotic differences of 150 mosm l^{-1} (Heisey et al. 1962). Accordingly, the model epithelium can transport against larger osmotic gradients when these are implemented by NaCl instead of mannitol. In the language of irreversible thermodynamics the reflection coefficient for NaCl, σ_{NaCl}, is lower than one. On the model σ_{NaCl} is 0.6, equal to 65 mosm l^{-1} divided by 110 mosm l^{-1}. This agrees with observations in the proximal tubule where σ_{NaCl} is between 0.5 and 0.8, for references see Zeuthen (1992).

10
Summary and Conclusions

There is good evidence that cotransporters of the symport type behave as molecular water pumps, in which a water flux is coupled to the substrate fluxes. The free energy stored in the substrate gradients is utilized, by a mechanism within the protein, for the transport of water. Accordingly, the water flux is secondary active and can proceed uphill against the water chemical potential difference. The effect has been recognized in all symports studied so far (Table 1). It has been studied in details for the K^+/Cl^- cotransporter in the choroid plexus epithelium, the H^+/lactate cotransporter in the retinal pigment epithelium, the intestinal Na^+/glucose cotransporter (SGLT1) and the renal Na^+/dicarboxylate cotransporter both expressed in *Xenopus* oocytes. The generality of the phenomenon among symports with widely different primary structures suggests that the property of molecular water pumps derives from a pattern of conformational changes common for this type of membrane proteins.

Most of the data on molecular water pumps are derived from fluxes initiated by rapid changes in the composition of the external solution. There was no experimental evidence for unstirred layers in such experiments, in accordance with theoretical evaluations. Even the experimental introduction of unstirred layers did not lead to any measurable water fluxes.

The majority of the experimental data supports a molecular model where water is cotransported: A well defined number of water molecules act as a substrate on equal footing with the non-aqueous substrates. The ratio of any two of the fluxes is constant, given by the properties of the protein, and is independent of the driving forces or other external parameters.

The detailed mechanism behind the molecular water pumps is as yet unknown. It is, however, possible to combine well established phenomena for enzymes into a working model. For example, uptake and release of water is associated with conformational changes during enzymatic action; a specific sequence of allosteric conformations in a membrane bound enzyme would give rise to vectorial transport of water across the membrane.

In addition to their recognized functions, cotransporters have the additional property of water channels. Compared to aquaporins, the unitary water permeability is about two orders of magnitude lower. It is suggested that the water permeability is determined from chemical associations between the water molecule and sites within the pore, probably in the form of hydrogen-bonds. The existence of a passive water permeability suggests an alternative model for the molecular water pump: The water flux couples to the flux of non-aqueous substrates in a hyperosmolar compartment within the protein.

Molecular water pumps allow cellular water homeostasis to be viewed as a balance between pumps and leaks. This enables cells to maintain their intracellular osmolarity despite external variations. Molecular water pumps could be relevant for a wide range of physiological functions, from volume regulation in contractile vacuoles in amoeba to phloem transport in plants (Zeuthen 1992, 1996). They could be important building blocks in a general model for vectorial water transport across epithelia. A simplified model of a leaky epithelium incorporating $K^+/Cl^-/H_2O$ and Na^+/glucose/H_2O cotransport in combination with channels and primary active transport gives good quantitative predictions of several properties. In particular of how epithelial cell layers can transport water uphill.

Acknowledgements. Useful discussions with M.Sc. Nanna Boxenbaum and financial support from The Danish Medical Research Council and The Lundbeck Foundation are gratefully acknowledged.

References

Agre P, Bonhivers M, Borgnia MJ (1998) The aquaporins, blueprints for cellular plumbing systems. J Biol Chem 273:14659–14662

Appelboom JWTh, Brodsky WA, Tuttle WS, Diamond I (1958) The freezing point depression of mammalian tissues after sudden heating in boiling distilled water. J Gen Physiol 41:1153–1169

Aull F (1981) Potassium chloride cotransport in steady-state ascites tumor cells. Biochim Biophys Acta 643:339–345

Barry PH, Diamond JM (1984) Effects of unstirred layers on membrane phenomena. Physiol Rev 64:763–872

Baslow MH (1998) Function of the N-acetyl-L-histidine system in the vertebrate eye: Evidence in support of a role as a molecular water pump. J Mol Neurosci 10(3) 193–208

Baslow MH (1999) The existence of molecular water pumps in the nervous system: a review of the evidence. Neurochem Int 34:77–90

Baum M, Berry CA (1984) Evidence for neutral transcellular NaCl transport and neutral basolateral chloride exit in the rabbit proximal convoluted tubule. J Clin Invest 74:205–211

Bianchini L, Fossat B, Porthe-Nibelle J, Ellory JC, Lahlou B (1988) Effects of hypoosmotic shock on ion fluxes in isolated trout hepatocytes. J Exp Biol 137:303–318

Boulpaep EL, Maunsbach AB, Tripathi S, et al (1993) Mechanisms of isosmotic water transport in leaky epithelia: Consensus and inconsistencies. In: Ussing HH, Fischbarg J, Sten-Knudsen O, et al (eds) Isotonic Transport in Leaky Epithelia. Munksgaard, Copenhagen, pp 53–67

Bulone D, Donato ID, Palma-Vittorelli MB, Palma MU (1991) Density, structural lifetime, and entropy of H-bond cages promoted by monohydric alcohols in normal and supercooled water. J Chem Phys 94:6816–6826

Bundgaard M, Zeuthen T (1982) Structure of Necturus gallbladder epithelium during transport at low external osmolarities. J Membr Biol 68:97–105

Chen X-Z, Coady MJ, Jackson F, Berteloot A, Lapointe J-Y (1995) Thermodynamic determination of the Na+: glucose coupling ratio for the human SGLT1 cotransporter. Biophys J 69:2405–2414

Cherksey BD, Zeuthen T (1987) A membrane protein with a K^+ and a Cl^- channel. Journal of Physiology 387:33P

Chrispeels MJ, Maurel C (1994) Aquaporins: The molecular basis of facilitated water movement through living plant cells? Plant Physiol 105:9–13

Colombo MF, Rau DC, Parsegian VA (1992) Protein Solvation in Allosteric Regulation: A Water Effect on Hemoglobin. Science 256:655–659

Conway EJ, Geoghegan H (1955) Molecular concentration of kidney cortex slices. J Physiol 130:438–445

Conway EJ, Geoghegan H, McCormack JI (1955) Autolytic changes at zero centigrade in ground mammalian tissues. J Physiol 130:427–437

Cotton CU, Reuss L (1989) Measurement of the effective thickness of the mucosal unstirred layer in *Necturus* gallbladder epithelium. J Gen Physiol. 93:631–647

Cotton CU, Reuss L (1991) Effects of changes in mucosal solution Cl⁻ or K⁺ concentration on cell water volume of *Necturus* gallbladder epithelium. J Gen Physiol. 97:667–686

Cotton CU, Weinstein AM, Reuss L (1989) Osmotic water permeability of necturus gallbladder epithelium. J Gen Physiol 93:649–679

Curran PF, Macintosh JR (1962) A model system for biological water transport. Nature 193:347–348

Dainty J, House CR (1966) Unstirred layers in frog skin. J Physiol 182:66–78

Diamond JM (1979) Osmotic water flow in leaky epithelia. J Membr Biol 51:195–216

Doyle DA, Cabral JM, Pfuetzner RA, Kou A, Gulbis JM, Cohen SL, Chait BT, MacKinnon R (1998) The structure of the potassium channel: Molecular basis of K+ conduction and selectivity. Science 280:69–77

Eskandari S, Wright EM, Kreman M., Starace DM, Zampighi GA (1998) Structural analysis of cloned plasma membrane proteins by freeze-fracture electron microscopy. Proc Natl Acad Sci USA 95:11235–11240

Eveloff J, Warnock DG (1987) K-Cl transport systems in rabbit renal basolateral membrane vesicles. Am J Physiol 252:F883–F889

Fischbarg J, Kuang K, Vera JC, Arant S, Silverstein SC, Loike J, Rosen OM (1990) Glucose Transporters Serve as Water Channels. Proc Natl Acad Sci USA 87:3244–3247

Frambach DA, Weiter JJ, Adler AJ (1985) A photogrammetric method to measure fluid movement across isolated frog retinal pigment epithelium. Biophys J 47:547–552

Glasstone S, Laidler KJ, Eyring H (1941) The Theory of Rate Processes. McGraw-Hill Book Company, Inc, New York and London

Green R, Giebisch G, Unwin R, Weinstein AM (1991) Coupled water transport by rat proximal tubule. Am J Physiol 261:F1046–F1054

Greger R, Schlatter E (1983) Properties of the basolateral membrane of the cortical thick ascending limb of Henle's loop of rabbit kidney. Pflüg Arch 396:325–334

Guggino WB (1986) Functional heterogeneity in the early distal tubule of the Amphiuma kidney: evidence for two modes of Cl⁻ and K⁺ transport across the basolateral cell membrane. Am J Physiol 250:F430–F440

Halm DR, Krasny EJ, Frizzell RA (1985) Electrophysiology of flounder intestinal mucosa. I. Conductance properties of the cellular and paracellular pathways. J Gen Physiol 85:843–864

Hamann S, La Cour M, Lui GM, Zeuthen T (1996) Transport of protons lactate and water in cultured fetal human RPE. Invest Ophthalmol Vis Sci 37:1109

Hamann S, Zeuthen T, La Cour M, Nagelhus EA, Ottersen OP, Agre P, Nielsen S (1998) Aquaporins in complex tissues: distribution of aquaporins 1–5 in human and rat eye. Am J Physiol 274:C1332–C1345

Hasegawa H, Skack W, Baker O, Calayag MC, Lingappa V, Verkman AS (1992) A multifunctional aqueous channel formed by CFTR. Science 258:1477–1479

Heath H, Kang SS, Philoppou D (1975) Glucose, glucose-6-phosphate, lactate and pyruvate content of the retina, blood and liver of streptozotocin-diabetic rats fed sucrose- or starch-rich diets. Diabetologica 11:57–62

Hediger MA, Coady MJ, Ikeda TS, Wright EM (1987) Expression cloning and cDNA sequencing of the Na$^+$/glucose co-transporter. Nature 330:379–381

Heisey SR, Held D, Pappenheimer JR (1962) Bulk flow and diffusion in the cerebrospinal fluid system of the goat. Am J Physiol 203:775–781

Hill A (1980) Salt-water coupling in leaky epithelia. J Membr Biol 56:177–182

Hill BS, Hill AE (1978) Fluid transfer by necturus gall bladder epithelium as a function of osmolarity. Proc R Soc Lond 200:151–162

Hille B (1992) Ionic Channels of Excitable Membranes. Sinauer Associates Inc., Sunderland, Massachusetts

House CR (1974) Water transport in cells and tissues. Edward Arnold, London

Hughes BA, Miller SS, Machen TE (1984) Effect of cyclic AMP on fluid absorption across the frog retinal pigment epithelium. Measurements in the open circuit state. J Gen Physiol 83:875–899

Ikonomov O, Simon MX, Frömter E (1985) Electrophysiological studies on lateral intercellular spaces of *Necturus* gallbladder epithelium. Pflüg Arch 403:301–307

Ishibashi K, Kuwahara M, Gu Y, Kageyama Y, Tohsaka A, Suzuki F, Marumo F, Sasaki S (1997) Cloning and functional expression of a new water channel abundantly expressed in the testis permeable to water, glycerol, and urea. J Biol Chem. 272:20782–20786

Kornblatt JA (1998) The water channel of cytochrome c oxidase: inferences from inhibitor studies. Biophysical Journal 75:3127–3134

Kornblatt JA, Hoa GHB (1990) A nontraditional role for water in the cytochrome c oxidase reaction. Biochem 29:9370–9376

Kornblatt JA, Kornblatt MJ, Rajotte I, Hoa GHB, Kahn PC (1998) Thermodynamic volume cycles for electron transfer in the cytochrome c oxidase and for the binding of cytochrome c to cytochrome c oxidase. Biophys J 75:435–444

Kramhøft B, Lambert IH, Hofmann EK, Jørgensen F (1986) Activation of Cl dependent K transport in Ehrlich ascites tumor cells. Am J Physiol 251:C369–C379

Kregonow FM (1971) The response of duck erythrocytes to nonhemolytic hypotonic media. J Gen Physiol 58:372–395

Kuriyama H, Kawamoto S, Ishida N, Ohno I, Mita S, Matsuzawa Y, Matsubara K, Okubo K (1997) Molecular cloning and expression of a novel human aquaporin from adipose tisue with glycerol permeability. Biochem Biophys Res Com 241:53–58

laCour M, Lin H, Kenyon E, Miller SS (1994) Lactate transport in freshly isolated human fetal retinal pigment epithelium. Invest Ophthalmol Vis Sci 35:434–440

Larson M, Spring KR (1984) Volume regulation by necturus gallbladder: basolateral KCl exit. J Membr Biol 81:219–232

Latorre R, Miller C (1983) Conduction and selectivity in potassium channels. J Membr Biol 71:11–30

Lauf PK, Bauer J, Adragna NC, Fujise H, Martin A, Zade-Oppen M, Ryu KH, Delpire E (1992) Erythrocyte K-Cl cotransport: properties and regulation. Am J Physiol 263:C917–C932

Loike JD, Hickman S, Kuang K, Xu M, Cao L, Vera JC, Silverstein SC, Fischbarg J (1996) Sodium-glucose cotransporters display sodium- and phlorizin-dependent water permeability. Am J Physiol 271:C1774–C1779

Loo DDF, Wright EM, Meinild A-K, Klaerke DA, Zeuthen T (1999a) commentary on "Epithelial Fluid Transport – A century of investigation". News Physiol Sci 14:98–100

Loo DDF, Zeuthen T, Chandy G, Wright EM (1996) Cotransport of water by the Na⁺/glucose cotransporter. Proc Natl Acad Sci USA 93:

Loo DF, Hirayama BA, Meinild A-K, Chandy G, Zeuthen T, Wright E (1999b) Passive water and ion transport by cotransporters. J Physiol, in press

Lubin M (1963) Cell potassium and the regulation of protein synthesis. In: Hoffman JF (ed) The Cellular Functions of Membrane Transport. Prentice-Hall Inc., Englewood Cliffs, New Jersey, pp 193–211

Ludwig C (1861) Lehrbuch der Physiologie des Menschen. C.F. Wintersche Verlagshandlung, Leipzig und Heidelberg

Ma T, Frigeri A, Hasegawa H, Verkman AS (1994) Cloning of a water channel homolog expressed in brain meningeal cells and kidney collecting duct that functions as a stilbene-sensitive glycerol transporter. J Biol Chem 269:21845–21849

Mackenzie B, Loo DDF, Wright EM (1998) Relationship between Na+/glucose cotransporter (SGLT1) currents and fluxes. J Membr Biol 162:101–106

Maffly RH, Leaf A (1959) The potential of water in mammalian tissues. J Gen Physiol 42:1257–1275

Marmor MF (1989) Mechanisms of normal retinal adhesion. In: Ryan SJ, Schachat AP, Murphy RB, et al (eds) Retina. Mosby, St. Louis, USA, pp 71–87

Maurel C, Reizer J, Schroeder JI, Chrispeels MJ, Saier Jr MH (1994) Functional Characterization of the *Escherichia coli* Glycerol Facilitator, GlpF, in *Xenopus* Oocytes. J Biol Chem 269:11869–11872

Meinild A-K, Klaerke DA, Loo DDF, Wright EM, Zeuthen T (1998a) The human Na⁺/Glucose cotransporter is a molecular water pump. J Physiol 508.1:15–21

Meinild A-K, Klaerke DA, Zeuthen T (1998b) Bidirectional water fluxes and specificity for small hydrophilic molecules in aquaporins 0 to 5. J Biol Chem 273:32446–32451

Meinild A-K, Loo DFF, Pajor A, Zeuthen T, Wright EM (2000) Water transport by the renal Na+/Dicarboxylate Cotransporter. Am J Physiol (in press)

Mitchell P (1957) A general theory of membrane transport from studies of bacteria. Nature 180:134–136

Mitchell P (1990) Osmochemistry of solute translocation. Res Microbiol 141:286–289

Nikaido H, Saier MH, Jr. (1992) Transport proteins in bacteria: Common themes in their design. Science 258:936–942

Parent L, Supplison S, Loo DDF, Wright EM (1992) Electrogenic properties of the cloned Na+/Glucose Cotransporter. J Membr Biol 125:49–62

Parsegian A, Rau D, Zimmerberg J (1993) Structural transitions induced by osmotic stress. In: Leopold AC (ed) Membranes, Metabolism, and Dry Organisms. Comstock Publishing Associates, London, pp 306–317

Parsons DS, Wingate DL (1961) The effect of osmotic gradients on fluid transfer across rat intestine in vitro. Biochim Biophys Acta 46:170–183

Patlak CS, Goldstein DA, Hoffman JF (1963) The flow of solute and solvent across a two-membrane system. J Theor Biol 3:420–442

Pedley TJ (1978) The development of osmotic flow through an unstirred layer. J Theor Biol 70:427–447

Pedley TJ (1983) Calculation of unstirred layer thickness in membrane transport experiments: a survey. Quart Rev Biophys 16:115–150

Pedley TJ, Fischbarg J (1980) Unstirred layer effects in osmotic water flow across gallbladder epithelium. J Membr Biol. 54:89–102

Perry PB, O'Neill WC (1993) Swelling-activated K fluxes in vascular endothelial cells: volume regulation via K-Cl cotransport and K channels. Am J Physiol 265:C763–C769

Persson B-E, Spring KR (1982) Gallbladder epithelial cell hydraulic water permeability and volume regulation. J Gen Physiol 79:481–505

Preston GM, Carroll TP, Guggino WB, Agre P (1992) Appearance of water channels in zenopus oocytes expressing red cell CHIP28 protein. Science 256:385–389

Qian M, Haser R, Payan F (1995) Carbohydrate Binding Sites in a Pancreatic α-Amylase-Substrate Complex, Derived from X-ray Structure Analysis at 2.1 Å Resolution. Prot Sci 4:747–755

Rand RP, Fuller NL (1992) Water as an inhibiting ligand in yeast hexokinase. Biophys J 61:A345

Reid EW (1892) Report on experiments upon "absorbtion without osmosis". Brit Med J 1:323–326

Reid EW (1901) Transport of fluid by certain epithelia. J Physiol 26:436–444

Reuss L (1983) Basolateral KCl co-transport in a NaCl-absorbing epithelium. Nature 305:723–726

Reuss L (1985) Changes in cell volume measured with an electrophysiologic technique. Proc Natl Acad Sci 82:6014–6018

Sasaki Sea (1988) KCl co-transport across the basolateral membrane of rabbit renal proximal straight tubules. J Clin Invest. 81:194–199

Schnermann J, Chou C-L, Ma T, Traynor T, Knepper MA, Verkman AS (1998) Defective proximal tubular fluid reabsorbtion in transgenic aquaporin-1 null mice. Proc Natl Acad Sci USA 95:9660–9664

Schreiber R, Greger R, Nitschke R, Kunzelmann K (1997) Cystic fibrosis transmembrane conductance regulator activates water condutance in Xenopus oocytes. Pflüg Arch 434:841–847

Shindo T, Spring KR (1981) Chloride movement across the basolateral membrane of proximal tubule cells. J Membr Biol. 58:35–42

Spring KR (1998) Routes and mechanism of fluid transport by epithelia. Ann Rev Physiol 60:105–119

Spring KR (1999) Epithelial Fluid Transport - A Century of Investigation. News Physiol Sci 14: 94–97

Steinbach HB (1962) The prevalence of K. Perspect Biol Med 5:338–355

Steitz TA, Shoham M, Bennett WS, Jr. (1981) Structural dynamics of yeast hexokinase during catalysis. Phil Trans R Soc Lond 293:43–52

Tsukaguchi H, Shayakul C, Berfer UV, Mackenzie B, Devidas S, Guggino WB, Van-Hoek AN, Hediger MA (1998) Molecular Characterization of a Broad Selectivity Neutral Solute Channel. J Biol Biochem 273:24737–24743

Turk E, Kerner CJ, Lostao MP, Wright EM (1996) Membrane Topology of the Human Na$^+$/Glucose Cotransporter SGLT1. J Biol Chem 271:1925–1934

White TW, Bruzzone R, Goocenough DA, Paul DL (1992) Mouse Cx50, a functional member of the connexin family of gap junction proteins, is the lens fiber protein MP70. Mol Biol Cell 3:711–720

Whittembury G, Hill BS (1982) Fluid reabsorption by Necturus proximal tubule perfused with solutions of normal and reduced osmolarity. Proc Roy Soc 215:411–431

Wright EM, Loo DDF, Panayotova-Heiermann M, Lostao MP, Hirayama BH, Mack-
     enzie B, Boorer K, Zampighi G (1994) 'Active' sugar transprot in eurkaryotes. J
     Exp Biol 196:197-212
Wright EM, Loo DDF, Turk E, Hirayama BA (1996) Sodium cotransporters. Curr
     Opin Cell Biol 8:468-473
Wright EM, Loo DDF, Panayotova-Heiermann M, Lostao MP, Hirayama BH, Tyrk E,
     Eskandari S, Lam JT (1998) Structure and function of the Na^+/glucose cotrans-
     porter. Acta Physiol Scand Suppl 643:257-264
Wright EM, Wiedner G, Rumrich G (1977) Fluid secretion by the frog choroid
     plexus. Exp Eye Res Suppl 25:149-155
Yang B, Verkman AS (1997) Water and glycerol permeabilities of aquaporin 1-5 and
     MIP determined quantitatively by expression of epitope-tagged constructs in
     Xenopus oocytes. J Biol Chem 272:16140-14146
Zampighi GA, Kreman M, Boorer KJ, Loo DDF, Bezanilla F, Chandy G, Hall JE,
     Wright EM (1995) A Method for Determining the Unitary Functional Capacity of
     Cloned Channels and Transporters Expressed in Xenopus laevis oocytes. J
     Membr Biol 148:65-78
Zampighi GA, Kunig N, Loo DDF (1997) Structure and function of cell-to-cell chan-
     nels purified from the lens and of hemichannels expresed in oocytes. In: Latorre
     R, Saez JC (eds) From Ion Channels to Cell-to-Cell Conversations. Plenum Press,
     pp 309-321
Zeidel ML, Albalak A, Grossman E, Carruthers A (1992) Role of glucose carrier in
     human erythrocyte water permeability. Biochem 31:589-596
Zeidel ML, Ambudkar SV, Smith BL, Agre P (1992) Reconstitution of functional
     water channels in liposomes containing purified red cell CHIP28 protein. Bio-
     chem 31:7436-7440
Zeuthen T (1981) Isotonic transport and intracellular osmolarity in the necturus
     gall-bladder epithelium. In: Ussing HH, Bindslev N, Lassen NA et al. (eds)
     Munksgaard,Copenhagen, Copenhagen, pp 313-331
Zeuthen T (1982) Relations between intracellular ion activities and extracellular
     osmolarity in necturus gallbladder epithelium. J Membr Biol 66:109-121
Zeuthen T (1983) Ion activities in the lateral intercellular spaces of gallbladder epi-
     thelium transporting at low external osmolarities. J Membr Biol. 76:113-122
Zeuthen T (1991b) Secondary active transport of water across ventricular cell mem-
     brane of choroid plexus epithelium of Necturus maculosus. J Physiol 444:153-173
Zeuthen T (1991a) Water permeability of ventricular cell membrane in choroid
     plexus epithelium from necturus maculosus. Journal of Physiology 444:133-151
Zeuthen T (1992) From contractile vacuole to leaky epithelia. Biophys Biochem Acta
     1113:229-258
Zeuthen T (1993) Low reflection coefficient for KCl in an epithelial membrane. In:
     Ussing HH, Fischbarg J, Sten-Knudsen O et al. (eds) Isotonic Transport in Leaky
     Epithelia. Munksgaard, Copenhagen, pp 298-307
Zeuthen T (1994) Cotransport of K^+, Cl^- and H_2O by membrane proteins from chor-
     oid plexus epithelium of Necturus maculosus. J Physiol 478:203-219
Zeuthen T (1995) Molecular mechanisms for passive and active transport of water.
     Int Rev Cyt 160:99-161
Zeuthen T (1996) Molecular Mechanisms of Water Transport. Springer, Berlin, R.G.
     Landes Company, Texas
Zeuthen T, Hamann S, La Cour M (1996) Cotransport of H^+, lactate and H_2O by
     membrane proteins in retinal pigment epithelium of bullfrog. J Physiol 497:3-17

Zeuthen T, Klaerke DA (1999) Transport of water and glycerol in aquaporin 3 is gated by H+. J Biol Chem 274: 21631–21636

Zeuthen T, Meinild A-K, Klaerke DA, Loo DDF, Wright EM, Belhage B, Litman T (1997) Water transport by the Na$^+$/glucose cotransporter under isotonic conditions. Biol Cell 89: 307–312

Zeuthen T, Stein WD (1994) Co-transport of salt and water in membrane proteins: Membrane proteins as osmotic engines. J Membr Biol 137:179–195

Zimmerberg J, Bezanilla F, Parsegian VA (1990) Solute inaccessible aqueous volume changes during opening of the potassium channel of the squid giant axon. Biophys J 57:1049–1064

Zimmerberg J, Parsegian VA (1986) Polymer inaccessible volume changes during opening and closing of a voltage-dependent ionic channel. Nature 323:36

Zimmerman SB, Harrison B (1987) Macromolecular crowding increases binding of DNA polymerase to DNA: An adaptive effect. Proc Natl Acad Sci 84:1871–1875

Zimmerman SB, Minton AP (1993) Macromolecular crowding: Biochemical, biophysical, and physiological consequences. Ann Rev Biophys Biomol Struct 22:27–65

Role of Lateral Intercellular Space and Sodium Recirculation for Isotonic Transport in Leaky Epithelia

E. H. Larsen, S. Nedergaard, and H. H. Ussing

August Krogh Institute, The University of Copenhagen, Universitetsparken 13, 2100 Copenhagen Ø, Denmark

Contens

1
Scope of the Review

Some theories of fluid transport by leaky epithelia assume a dominant role of the paracellular space for coupling of water and solute fluxes. One such theory is the Na^+-recirculation theory introduced for dealing with solute coupled fluid transport in secretory frog skin glands (Ussing and Eskesen 1989, Ussing et al. 1996) and absorbing small intestine (Ussing and Nedergaard 1993). In the present review we compare the previously formulated theories on paracellular solute-solvent coupling with the recirculation theory. We subsequently discuss the recirculation of sodium ions emphasizing on recent studies of toad small intestine (Nedergaard et al. 1999, Nedergaard 1999) with a quantitative analysis showing how cellular and paracellular fluxes of leaky epithelia can be separated. We present evidence for the paracellular space being the coupling compartment for ion and water fluxes in small intestine, and discuss equations that can be used for estimating the recirculation flux of Na^+. The discussion is extended by outlining a recent theoretical treatment of epithelial fluid transport energized by active solute pumps and recirculation of the driving solute (Larsen and Sørensen 1999, Larsen et al. 2000). Finally, the review contains – in the light of the recirculation theory – a discussion of some published experimental observations on other vertebrate absorbing epithelia of the leaky type.

2
General Considerations

2.1
Statement of the Problem

The types of epithelia that will be discussed utilize metabolic energy to absorb isotonic fluid in the absence of transepithelial ion concentration-, electrical potential-, osmotic-, and hydrostatic-pressure differences. Absorption of water depends on simultaneous absorption of NaCl with Na^+ submitted to active transport via Na^+/K^+-pumps, and can take place against a pre-established adverse osmotic gradient (Curran and Solomon 1957, Parsons and Wingate 1958, Windhager et al. 1958, Curran 1960, Diamond 1962a,b, 1964a,b, Green et al. 1991). A significant transepithelial electrical conductance is also shared by these epithelia. This leakiness to diffusible ions and electroneutral hydrophilic molecules has been ascribed unequivocally to the paracellular route (Boulpaep 1971, Frizzell and Schultz 1972, Frömter 1972, 1973 and 1986, Frömter et al. 1973, Halm et al. 1985a, Reuss 1991, Spring

1973a,b). The plasma membranes of the epithelial cells exhibit electrical conductances of the same order of magnitude as those of high-resistance epithelia (Boulpaep 1971, Frömter 1972, 1979, Frömter et al. 1971, Frömter, et al. 1973, Halm et al. 1985a,b, Reuss 1991, Reuss and Finn 1975b, Zeuthen 1977). Small intestine, gallbladder, and kidney proximal tubule are the classical leaky epithelia that have been used as experimental models in studies on the mechanism of isotonic transport. Although upper airway epithelia cannot really be classified as leaky (Boucher and Larsen 1988, Willumsen and Boucher 1989) they do transport salt and water in near isotonic proportions (Boucher 1994a,b, 1999) and may thus be considered among epithelia generating solute coupled water transport. Also amphibian skin carries out solute coupled water transport (Nielsen 1997), which appears to be of significance for rapid transcutaneous water uptake in dehydrated toads (Sullivan et al. 2000).

Briefly stated, the challenge is to understand within a common framework: (*i*) How metabolically energized transport of water and NaCl in isotonic proportions takes place with unmeasurable external driving forces. (*ii*) How leaky epithelia, of which some have a significant paracellular water permeability, transport water uphill. (*iii*) How anomalous solvent drag demonstrated in amphibian skin (Ussing 1966, Ussing and Johannsen 1969) can be reconciled with solute coupled water transport.

2.2
Models with Subepithelial Coupling Compartments

2.2.1
The Curran-Patlak and Diamond-Bossert Models

The type of models to be considered in this review assume coupling of solute and water flows in a macroscopic subcompartment that builds up osmotic and hydrostatic pressures above those of bathing solutions. Such a three compartment/two-membrane system first suggested by Curran (1960) and depicted in Fig. 1A, was studied experimentally by Curran and MacIntosh (1962) and theoretically by Patlak et al. (1963). By active transport across the 1st membrane a solute is accumulated in the middle compartment, and by osmosis water is drawn into this compartment through the 1st membrane with reflection coefficient of unity ($\sigma^{1,2} \cong 1$). By assuming that the 2nd membrane is freely permeable to both solute and water ($\sigma^{2,3} = 0$), fluid would be forced into the inner compartment by a hydrostatic pressure difference between the middle and the inner compartment. In a subsequent theoretical analysis of such a two-membrane system, Patlak et al. (1963) concluded that if $\sigma^{1,2} \neq \sigma^{2,3}$, the system has capacity of producing an anisotonic transportate giving the „*appearance that solvent is pumped, whereas the solvent movement is actually due to passive forces*" (*loc. cit.* p 434). With $\sigma^{2,3} = 0$, as originally discussed in Curran (1960), and with well stirred compartments, they showed that the transportate would always be hypertonic, but for $0 < \sigma^{2,3} < \sigma^{1,2}$, the inwardly directed absorbate (left to right in Fig. 1A) would be either hypotonic, isotonic or hypertonic. Furthermore, with fixed values of reflection coefficients truly isotonic transport would occur at one rate of fluid movement, only. If, as observed in laboratory experiments, isotonicity of the absorbed fluid is achieved for a large range of transport rates, according to the possibilities offered by their analysis, this could be achieved by controlling reflection coefficients.

In a study of fluid transport by gallbladder, Whitlock and Wheeler (1964) suggested that the middle compartment of the Curran-Patlak model was the lateral intercellular space, and with these spaces being like long narrow channels (Tormey and Diamond 1967), it was indicated that the middle compartment would not be well stirred. In the subsequently formulated standing gradient theory (Diamond and Bossert 1967, 1968) isotonicity was achieved by assuming solute pumps to be concentrated upstream on the lateral membranes. Together with certain geometrical properties of the lateral space it was assumed that both solute and water enter the lateral space via adjacent epithelial cells and that water and ions by diffusion are becoming in equilibrium with the bathing solutions by reaching isotonicity before leaving the space (see Fig. 1B). Analysis of the standing gradient theory was executed by mathematical modelling of *Necturus* proximal tubule with ion permeable tight junctions and evenly distributed pumps along the lateral space (Sackin and Boulpaep 1975). It was concluded that with these plausible modifications (see below) it would be virtually impossible to achieve an isotonic transportate by relying on diffusion equilibrium in lateral spaces of relevant physical dimensions. Also other theoretical treatments of the gradient theory, or modifications thereof, concluded that a standing gradient mechanism depends critically on boundary conditions as defined by assumed values of membrane-parameters (permeabilities, length and width of lateral space) with unstable fluid flows in neighbourhood of acceptable mathematical solutions (Weinstein and Stephenson 1981a, Rubinstein and Segel 1983).

A

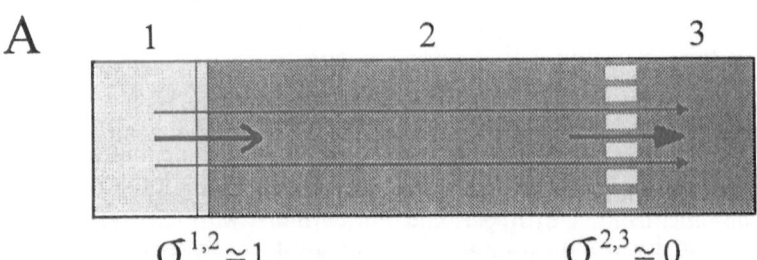

$$\sigma^{1,2} \simeq 1 \qquad \sigma^{2,3} \simeq 0$$

B

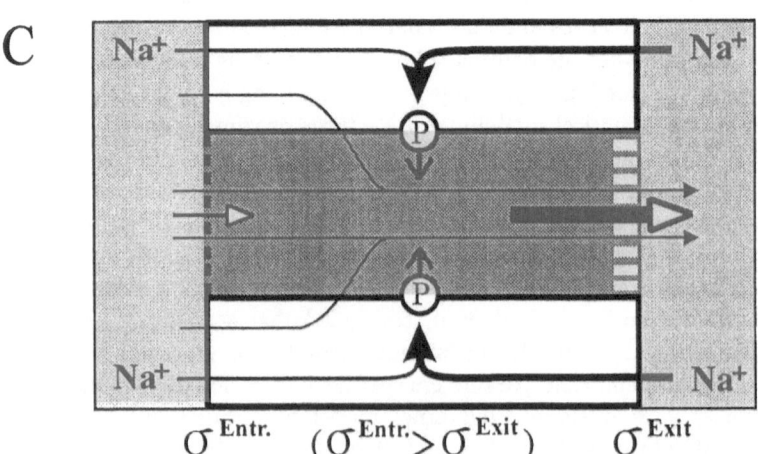

Solute
pumps
→

Water
flow
→

C

Na⁺ — ... — Na⁺

Na⁺ — ... — Na⁺

$$\sigma^{Entr.} \quad (\sigma^{Entr.} > \sigma^{Exit}) \quad \sigma^{Exit}$$

Ⓟ→ Na/K-pump flux

➤ Downhill Na⁺ flux

▷ Convective Na⁺ flux

→ Water flow

2.2.2
Models with Ion Recirculation

It should be noted, in the Curran-Patlak and Diamond-Bossert models – as well as in all subsequent quantitative descriptions derived from the ideas originally formulated by Curran (1960), *e.g.*, Weinstein and Stephenson (1981b), Tripathi and Boulpaep (1989), Spring (1998, 1999) – metabolic energy is invested only in the generation of an hypertonic and hyperbaric solution within the epithelium. The subsequent adjustment of the tonicity of the transportate is postulated to be a purely dissipative process. The Na^+ recirculation theory is different from the models depicted in Figs. 1A and B in postulating a dual role of cell metabolism for production of a truly isotonic transportate (see Fig. 1C). Firstly, by increasing the tonicity of lateral intercellular space, Na^+/K^+-pumps are driving water into the epithelium. Secondly, by cellular reuptake of Na^+ (together with K^+ and Cl^-) the Na^+/K^+-pumps return the solutes to the lateral space, such that the tonicity of the *net*

Fig. 1. Three different models of coupling between solute and water flows in a subcompartment within the epithelium. A: In the two-membrane model by Curran (1960) and Curran and MacIntosch (1962) a solute is accumulated by active transport in the middle compartment. By osmosis water is taken up through the 1st membrane with reflection coefficient, $\sigma^{1, 2} \approx 1$. By bulk flow, the fluid enters the inner compartment through a barrier with reflection coefficient, $\sigma^{2, 3} \approx 0$. B: The standing gradient theory of Diamond and Bossert (1967) assumes that isotonicity of the transportate is achieved by dissipation of local concentration gradients within the middle compartment, *i.e.* the lateral intercellular space. The active solute pumps are located upstream on the membranes lining this space. Also water has to enter this space via transport through adjacent cells. C: According to the Na^+ recirculation theory metabolic energy is invested both in the formation of hypertonic intercellular space fluid, and in the subsequent adjustment of the tonicity of the transportate (Nedergaard et al. 1999). Both of these works are performed by Na^+/K^+-pumps on the membranes lining the lateral intercellular space. Water may enter the lateral intercellular space through tight junctions and/or through cells. This may vary among different types of absorbing epithelia, and is not of significance for the principal mechanism (Larsen et al. 2000). From the lateral intercellular space the fluid proceeds to the inner compartment through the interspace basement membrane forced by the hydrostatic pressure difference between interspace and inner compartment, *i.e.*, $\sigma^{Entrance} > \sigma^{Exit}$. Sodium ions enter the space via cells (active transport) and through tight junction (convection). The sum of these components constitutes the Na^+ flux exiting the lateral intercellular space. A major fraction of this flux is recirculated through cells. This is indicated by the relative thickness of arrows

transportate is adjusted to be similar to that of the bathing solutions. In our original paper, it was suggested that water is being transported across the intestinal epithelium via tight junction through the paracellular pathway (Nedergaard et al. 1999). But with recirculation, isotonic transport can also be achieved with no hydraulic conductance of tight junction provided water channels are expressed in apical and lateral membranes so that water can take a translateral route (Larsen et al. 2000). Whatever pathway water takes for entering the lateral space – which might well vary among epithelia from different tissues – exit of water across the interspace basement membrane provides the condition for convection-diffusion fluxes of diffusible electrolytes and other hydrophilic molecules that have entered the lateral space either via tight junctions or cells. In principle, subcompartments within the epithelium other than the lateral intercellular space, like basal infoldings with, if any, associated trafficking intracellular vesicles, might constitute additional coupling compartments.

Solute recycling as a means of regulating the tonicity of an epithelial absorbate was first discussed by Berridge and Gupta (1967) and further elaborated upon by Gupta and Hall (1981) in a study of rectal papillae of the blowfly *Calliphora erythrocephala*. It was hypothesized that ions were being reabsorbed along the lateral membranes of the winding intercellular labyrinths with K^+ as primary solute and – depending on the final tonicity of the transportate – with additional H_2O recycling via the cells between serosal solution and lateral space. Our recirculation theory is different from the one suggested for the insect, (1) Na^+, not K^+, plays the major role in regulating the transportate's tonicity, (2) the driving ion species (*i.e.* the sodium ion of our model) is not taken up from the paracellular compartment, but rather is pumped into it from the cells, (3) water recycling is not necessary.

2.3
Evidence for an Extracellular Subepithelial Coupling Compartment

Since the first notion (Whitlock and Wheeler 1964) it has been a debated and controversial question whether the lateral intercellular space does constitute a transit compartment for transepithelial solute and water fluxes of functional significance for isotonic transport. Here we summarize several types of approaches that have been used in previous investigations for examining this point. Some are being dealt with in more detail in subsequent sections in which our own investigations are discussed.

2.3.1
Localization of Sodium Pumps

Studies on the distribution of $^3H^+$-labelled ouabain have revealed relatively dense concentration of Na^+/K^+-pumps on membranes lining lateral intercellular spaces of leaky epithelia (Stirling 1972, Mills and DiBona 1978, DiBona and Mills 1979) indicating that a major fraction of the actively absorbed Na^+ passes the lateral space before entering the serosal bath. There seems to be at least one exception to this notion. In *Ambystoma* kidney proximal tubule electron microscopical localization of antibodies raised against the Na^+/K^+-ATPase indicated high pump expression on membranes lining basal infoldings (Maunsbach and Boulpaep 1991). As these membranes constitute about 75% of the total baso-lateral membrane area (Maunsbach and Boulpaep 1984) it was suggested that the well-developed basal infoldings of the nephron segment of this species constitute a quantitatively important transit pathway for reabsorbed Na^+. Taken together with measured high water permeabilities of the apical and basolateral plasma membranes, evidence was provided that in this species the basolateral extracellular labyrinths constitute a major solute-solvent coupling compartment (Boulpaep et al. 1993).

2.3.2
Volume Changes of the Lateral Intercellular Space

The compliance of the lateral intercellular space has been estimated for two epithelia, *i.e.*, *Necturus* gallbladder (Spring and Hope 1978) and MDCK cells (Timbs and Spring 1996) and the values differ by a factor of about five. Since the yielding of the epithelial plasma membranes determines how much the lateral intercellular space becomes deformed at particular hydrostatic pressure of the space, volume changes of the space in relation to imposed fluid flow has to be interpreted with caution.

Microscopic studies of fixed and sectioned transporting leaky epithelia have indicated expansion and reduction of lateral intercellular space with stimulation and inhibition, respectively, of transepithelial fluid flow (Witlock and Wheeler 1964, Kaye et al. 1966; Tormey and Diamond 1967, Atisook et al. 1990). However, preparative methods of electron microscopy were suggested to induce cell shrinkage with secondary intercellular space expansions (Frederiksen and Leyssac 1974). In a pioneering work Spring and Hope (1978, 1979) developed a video-microscopical method for measuring size and shape of cells and lateral intercellular spaces while the epithelium is transporting fluid. By varying the transepithelial hydrostatic pressure difference, the lateral space of the *Necturus* gallbladder was shown to constitute highly compliant elements, and when mucosal NaCl was replaced by osmotically equivalent amount of sucrose the lateral space and the cells reversibly lost volume to about 16% and 76%, respectively, of their control values. Within the experimental error the cell volume loss corresponded to an isosmotic emptying of intracellular NaCl, and from the initial rate of cell volume changes it was calculated that the rate of exchange of the intracellular NaCl-pool was of a magnitude similar to the active transepithelial flux of the salt under control conditions. Zeuthen 1983 stimulated transepithelial fluid transport in unilateral preparations of frog gallbladder by diluting the luminal solution to 62 ± 1.7 mOsm. Still the absorbate was isosmotic, 62 ± 1.1 mOsm. This stimulation of isotonic transepithelial fluid transport resulted in widening of the lateral space examined *in vitro* with Nomarski optics.

2.3.3
Paracellular versus Transcellular Water Permeability

A number of techniques have been developed for estimating overall transepithelial and transcellular water permeability (hydraulic conductance), respectively, of the leaky epithelium. In such studies it is necessary to perturb the external driving force for water movement across the epithelium, preferentially by exposing either luminal or serosal side to an anisosmotic solution. By recording the resulting transepithelial water flux, in principle, the transepithelial water permeability can be calculated. Repeating the measurements with pharmacologically blocked or genetically eliminated transcellular water permeability, if possible, the paracellular water permeability can then be estimated. Another way of estimating the cellular permeability is to record the initial rate of cell volume change in response to fast unilateral osmotic perturbations. This technique has the advantage that the water permeability of apical and basolateral membranes can be estimated separately. Cell volume changes have been monitored either by videomicroscopical examinations (Spring and Hope 1978), or by recording with ion-selective microelectrodes intracellular activity of a non-permeant ionprobe (Zeuthen 1982, Reuss 1985). A common difficulty of all methods mentioned is that unless the imposed osmotic gradient is being corrected for unstirred layers adjacent to the epithelial cells, the water permeability might be grossly underestimated. Furthermore, the osmotic perturbation should be carried out using a solute with reflection coefficient of unity. If this cannot be done, the reflection coefficient of the solute of the two parallel pathways for water flow has to be estimated, which may impose another difficulty in the interpretation of measurements. Whittembury and Reuss (1992) reviewed the different methods with a discussion of their applications.

The hydraulic conductivity of the apical and basal membrane of the epithelial cells of *Necturus* gallbladder has been estimated to be, $1.0 \cdot 10^{-3}$ and $2.2 \cdot 10^{-3}$ $cm^3 \cdot s^{-1} cm^{-2} \cdot Osmol^{-1}$, respectively ($P_{osm}$ = 0.055 and 0.12 $cm \cdot s^{-1}$; Persson and Spring 1982). This study applied fast video-microscopy to cells undergoing volume changes in response to rapid switching between perfusion solutions of different osmolarity. High water permeabilities were also calculated from recording the activity of positively charged membrane impermeable cell volume markers during external osmotic perturbations (Zeuthen 1982, Reuss 1985, Cotton et al. 1989). Disregarding resistance to water transport through cytoplasm, with the above high permeabilities, the transmembrane osmotic concentration differences was calculated to be no more than 1–2 $mOsmol \cdot kg^{-1}$ required for driving water through the cells at the rate corresponding to the transepithelial fluid transport ($\sim 2 \cdot 10^{-6}$ $cm \cdot s^{-1}$, Persson and Spring 1982). This led to the hypothesis that in gallbladder water flows through cells. It was emphasized that the absorbate would never be truly isotonic, but just "near-isotonic". A similar hypothesis was forwarded for late proximal mammalian (rabbit, rat) nephron (Andreoli and Schafer 1978a, 1978b, 1979), whereas the transepithelial osmotic gradient required for driving water reabsorption in *Necturus* proximal tubule appears to be too large for escaping reliable measurements (discussed in Weinstein 1992). Contrasting the above conclusions, studies in Whittembury's laboratory indicated that the transcellular osmotic water permeability of rabbit proximal straight tubule is less than half of this segment's transepithelial water permeability (~ 30 and ~ 80 $cm \cdot s^{-1} \cdot Osmol^{-1}$, respectively), indicating a physiologically significant water permeability of tight junctions (Carpi-Medina and Whittembury 1988).

With the purpose of testing whether the water channel forming membrane protein, aquaporin-1 (AQP1, Nielsen and Agre 1995), constitutes the dominating route for transepithelial osmotic water reabsorption by proximal tubule, Schnermann et al. (1998) disrupted the AQP1-gene for studying osmotic water permeability of proximal tubule of [–/–] knockout mice. In wild type mice this gene is expressed in both apical and basolateral membrane. Proximal tubules of AQP1 knockout mice exhibited a decreased, by 78%, osmotic permeability (P_f) of the tubule wall. This was measured by recording transtubular water uptake of microperfused proximal tubules in the presence of a 50-mOsm osmotic gradient. Calculation showed that P_f = 0.15 ± 0.03 $cm \cdot s^{-1}$ in [+/+] mice, and P_f = 0.033 ± 0.005 $cm \cdot s^{-1}$ in [–/–] knockout mice. The remaining osmotic water permeability of [–/–] mice was suggested to be due to transport across the lipid portion of the membranes, and it was concluded that < 20% of water uptake by osmosis is paracellular (see also discussion by Agre 1998).

2.3.4
Transepithelial Convection

The classical study of rat proximal convoluted tubule by Frömter et al. (1973) indicated that significant components of reabsorbed sodium and chloride ions are due to paracellular convection. A significant convective flux of Cl⁻ was also indicated in a study of rabbit superficial proximal straight tubules (Andreoli et al. 1979). The reflection coefficients for Na^+ and Cl⁻ were estimated to, $\sigma_{Na} = 0.7$ and $\sigma_{Cl} = 0.5$, respectively (Frömter et al. 1973), but higher values have also been reported (Schafer et al. 1975, Hierholzer et al. 1980). Also the early studies of other leaky epithelia (amphibian proximal tubule: Boulpaep 1971, small intestine: Frizzell and Schultz 1972, gallbladder: Frömter 1972) provided evidence for relatively large ion conductances of tight junction, and this conclusion has been confirmed by all subsequent studies. The existence of low-resistance paracellular passage ways for ions diffusing down their electrochemical gradient would imply that tight junctions are furnished with pore forming proteins. Not until very recently was the first tight junction-protein with ion channel properties cloned, paracellin-1 (PCLN-1). PCLN-1 was shown to be permeable for Mg^{2+} and Ca^{2+} and identified as regulated pathway for paracellular reabsorption of Mg^{2+} in thick ascending limb of Henle (Simon et al. 1999). As this nephron segment is impermeable to water, indirectly PCLN-1 also provides the first example of an identified tight junction ion channel that seems to be water impermeable.

Whereas small diffusible ions can pass tight junction of leaky epithelia, the hypothesis that they are being transported across this structures in kidney by convection (Frömter et al. 1973; Andreoli et al. 1979) was questioned by Jacobson et al. 1982 in a study of Cl⁻ and HCO_3^- fluxes in rabbit proximal tubules. They could not detect consistent changes of the fluxes of these anions in response to an imposed osmotic water flow across rabbit proximal tubules. According to these authors this could be due either to near-unity reflection coefficients of NaCl and $NaHCO_3$, or because the induced water flow was confined to the cellular pathway. Spring and his colleagues have developed fluorescent microscopical techniques for measuring directly the concentration of small diffusible ions of the lateral intercellular space of MDCK cells. They estimated that about 20 % of the sodium ions pumped from the cells into the lateral space leaked back into the apical solution via tight junctions (Chatton and Spring 1995). This result is incompatible with the notion that sodium ions are submitted to inward convective transport across the tight junction membrane of MCDK cells grown to confluence.

The study of convective fluxes of inert molecules which cannot pass the cell membranes have been used as another, but still indirect, method of probing transjunctional water flows. With this method the probe is maintained at thermodynamic equilibrium across the epithelium so that a net transepithelial flux of the probe would indicate transport by convection. By choosing cell membrane impermeable probes, like sucrose[1], inulin, and dextrans, this method circumvents the principal problem of separating the solute fluxes through the different pathways of the structurally complex epithelium. Another method of separating paracellular and cellular fluxes shall be discussed in *Section* 3.1.2. In an investigation using sucrose as probe, Hill and Hill 1978 found that the concentration of sucrose of the fluid absorbed by *Necturus* gallbladder was almost as large as its concentration of the luminal fluid. Their calculations indicated that this might be brought about by convection if the entire transepithelial water uptake was extracellular. In a study of guinea-pig gallbladder Whittembury et al. (1980) demonstrated that the net flux of sucrose, inulin, and dextran was a linear function of the experimentally perturbed transepithelial water flux, and they suggested that water flows into the lateral spaces directly through *zonulae occludentes* with more than half of the physiological water flux being paracellular. In rat kidney proximal tubule fluxes of radiolabelled sucrose was studied with ^{3}H-inulin as volume marker (Whittembury et al. 1988, Whittembury et al. 1985). A fairly large variation of transtubular volume flow (J_V) was obtained by varying the luminal perfusion rate or by adding raffinose as impermeant osmolyte to the luminal perfusion solution. It was observed that sucrose reabsorption (J^S) normalized to the sucrose concentration (C_l^S) of the fluid collected downstream the perfused tubular segment (*i.e.*, J^S/C_l^S) increased very significantly, about 60 times, when J_V was increased from 0 to ~30 nl·cm^{-2}·s^{-1}. This result was analyzed on the basis of a model that included effects of unstirred layers on the diffusive fluxes through the tubular wall. Calculations indicated that an unstirred layer of several cm-thickness would be required to explain the strong dependence of J^S/C_l^S on J_V ruling out that pseudo-solvent drag (Barry and Diamond 1984) could explain the findings. By default, it was concluded that sucrose by convection is being transported through tight junctions. Recently this hypothesis was further investigated in gallbladder studies of unidirectional fluxes of radioactive dextrans with radii varying from 0.4–2.2 nm (Hill and Shachar-

---

[1] The presence of sucrases prevents this probe to be used in preparations of small intestine.

Hill 1993, Shachar-Hill and Hill 1993). Dextran-molecules of all sizes tested exhibited a highly significant net flux in luminal-to-serosal direction. The net-flux was a linear function of molecular radius with an intercept of $2 \cdot 10^{-6}$ cm·s^{-1} (zero probe radius), which was similar to the transepithelial linear fluid velocity estimated by other methods (Hill and Hill 1978, Larson and Spring 1983). Since the experiments were carried out in absence of any applied external driving forces, these findings were taken to indicate that dextran molecules by a metabolically driven convective flow are carried through tight junctions, *i.e.*, they concluded that tight junctions of *Necturus* gallbladder would be permeable to water.

Spring's laboratory recently introduced a new quantitative method for probing volume flows between epithelial cells (Kovbasnjuk et al. 1998). The slowly diffusing 70.000 M_r fluorescin dextran was loaded into the lateral intercellular space of low-resistance MDCK-cells grown to confluence. By illumination with low-intensity laser light, optical sectioning, and digital deblurring of confocal microscopic image stacks they were able to estimate convection induced concentration profiles along the lateral intercellular space from which volume flow curves were constructed. With no osmotic gradient imposed across the epithelium ('isotonic' conditions), as well as with hyperosmotic apical, or basal solution, the estimated volume flow at the level of tight junctions was zero. Even after cyclicAMP-stimulated transjunctional ion conductance the volume flow rate at the level of the junctions was estimated to be zero under isotonic conditions; with 50-mOsm hyperosmotic apical bath a small flux of water in the outward direction was detected. Thus according to this study with cultured cells, water enters the epithelium via the apical membrane and not via tight junctions. The authors further concluded that the flow of water out of the lateral space, estimated with some uncertainty, would be consistent with the major flux of water leaving the MDCK-cells through lateral membranes rather than through the plasma membrane facing the permeable support of the epithelium. They concluded, therefore, that water movement across the MDCK monolayer is not transjunctional, but translateral.

2.3.5
Osmotic and Ion Concentrations of Lateral Intercellular Fluid

Ikonomov et al. (1985) used ion sensitive microelectrodes for measuring electrolyte activities of the lateral intercellular space fluid of *Necturus* gallbladder and found that the transjunctional osmotic gradient could be no more than 1–2 mosmol·l^{-1}. A significantly larger concentration difference, corresponding to the lateral space concentration being ~4% above luminal solution and bath, was reported in an electrophysiological study of salamander proximal tubule (Sackin 1986). In a recent study of the Na^+ concentration of the lateral intercellular space of MDCK cells using the fluorescent dye, SPFO, the concentration of this ion was found to be about 15 mM above[2] that of bath (142 mM) with no detectable gradient along the space (Chatton and Spring 1995). Using the dye ABQ, within the experimental error the concentration of Cl^- was concluded to be about the same, or somewhat larger, than that of Na^+ (Xia et al. 1995). It therefore appears that the nominal osmotic concentration of the lateral intercellular space fluid of MDCK cells is about 40 $mosmol \cdot l^{-1}$ above that of the apical bath. These studies were conducted on cells grown to confluence on an impermeable support (glass coverslips), and it was discussed that the above-mentioned lateral ion concentrations of diffusible ions likely represent overestimates for cells grown on permeable support.

2.3.6
Summing up: Transjunctional or Translateral Water Transport?

Abundant expression of Na^+/K^+-pumps on the lateral membranes (*Section* 2.3.1) together with observed volume changes during fluid absorption (*Section* 2.3.2) provide strong evidence for both the cellular compartment and the lateral intercellular space constituting transit pathways for apically absorbed NaCl. A majority of studies concludes that water passes the lateral space during fluid absorption[3]. Three ways of transporting water into this space have been considered. Either water flows into the lateral space through the junctional membrane (transjunctional water transport), or via the apical plasma membrane through the membranes lining the lateral intercellular space (translateral water transport), or water flows via both of these routes. Direct studies of water flow along the lateral space of MDCK-cells, which is an experimental model of a distal segment of vertebrate nephron, showed that in this epithelium is the transjunctional water flux virtually zero (*Section* 2.3.4, Kovbasnjuk et al. 1998). The novel method used in this study requires simple geometry of the cultured epithelium and cannot, as yet, be applied to native epithelia of complicated three-dimensional geometry. Discussions of the water pathways in these epithelia may thus depend on

---

[2] This result was obtained in 24-mM bicarbonate/7% CO_2 buffered solutions. In HEPES buffered solutions, *lis*-[Na^+] was virtually identical with that of bath. A similar buffer system-dependent difference in lateral [Cl^-] was not found (Xia et al. 1995).

[3] *Ambystoma* proximal tubule appears to be an exception to this rule by having the major coupling compartment in infoldings of the basal membrane (Boulpaep et al. 1993; *Section* 2.3.1).

indirect studies. Both translateral and transjunctional water transport have been advocated in the literature discussed above. Presently, it is not known whether differences do exist among leaky epithelia of different organs and/or from different vertebrate species. Due to lack of direct evidence this question remains open. It should be emphasized, however, that all studies of flows of plasma-membrane impermeable hydrophilic molecules were interpreted (*Section* 2.3.4) with the assumption that experimentally verified transepithelial convective flows of the probe can be used for distinguishing transjunctional from translateral flow of water. Theoretically, at least, with respect to fluxes of tight junction-permeable solutes convective flows are to be obtained independent of the water flux being transjunctional or translateral (Larsen et al. 2000). Furthermore, within the framework of the Na^+-recirculation theory, it is not of fundamental significance for the mechanism of coupling between active transport of Na^+ and passive flow of water whether water enters the coupling compartment via tight junctions or cells.

3
Sodium Ion Recirculation in Small Intestine

Application of electrophysiological and molecular biologic techniques have indicated that leaky epithelia are furnished with cellular transport systems which would allow for both absorption and secretion of ions, independent of whether the normal function of the epithelium is associated with net fluid absorption or net fluid secretion. Thus, the quantitative analysis of transepithelial ion transport, beyond studying transepithelial net fluxes, also requires their unidirectional components being measured. Furthermore, the ratio of unidirectional paracellular fluxes provides important information on putative water flow between cells. Therefore, the aim of the studies to be discussed below has been to apply methods allowing for separation of cellular and paracellular unidirectional fluxes across unperturbed small intestine. We will show that the ratio of unidirectional paracellular fluxes provides important information on the relative significance of diffusive and convective solute flows out of the space. Based upon cellular and paracellular unidirectional fluxes we will also develop mathematical equations – referring to a simple compartment model – for estimating recirculation fluxes of the sodium ion.

3.1
Pre-Steady State Flux Ratio Analysis

3.1.1
Equations

Under certain conditions flux ratio analysis can provide the quantitative information we ask for, *i.e.*, the cellular and paracellular unidirectional fluxes in an unperturbed epithelium. With this method, radio-isotopes can be added to either side of the epithelium and the time course of isotopic-fluxes is followed until a steady-state is achieved. If the epithelium is in a physiological steady state, for a single pathway, such as the paracellular or the cellular route, and for any combination of transport mechanisms, the ratio of appearing isotopic-fluxes is independent of time (Sten-Knudsen and Ussing 1981):

$$\frac{M_{1 \to 2}(t)}{M_{2 \to 1}(t)} = \frac{C^{(1)}}{C^{(2)}} \exp\left(\frac{N_A \cdot W}{R \cdot T}\right) \tag{1}$$

The left hand side of the *Eq.* (1) is the ratio of unidirectional fluxes appearing at time *t* on the opposite side (the *trans* side) of the membrane to which the isotope was added (the *cis* side), *C* denotes concentration of the ion in compartment (*1*) and (*2*), respectively, N_A is Avogadro's number, *R* and *T* have their usual meaning, and *W* is the work done on the isotope molecule in moving it through the pathway. Notably, this work can be done by any combination of electric fields, drag associated with water flow, active transport, coupling to movement of other ions, etc., and the analysis can be applied to a multi-membrane system consisting of an arbitrary number of barriers provided the conditions governing the flux along the pathway, *e.g.*, electrical potentials, ion activities, diffusion coefficients, metabolic processes, etc., are time-invariant. The theoretical result given by *Eq.* (1) implies that the solute makes use of more than a single pathway if its flux ratio is time-variant at physiological steady state. With two Na$^+$ pathways, a cellular and a paracellular, the time course of the building up of steady state distribution of the isotope within the epithelium, are governed by equations of the form:

$$M_{ms}(t) = J_{ms}^{para} \cdot \left[1 - g(t)\right] + J_{ms}^{active} \cdot \left[1 - f(t)\right] \tag{2a}$$

$$M_{sm}(t) = J_{sm}^{para} \cdot \left[1 - g(t)\right] + J_{sm}^{active} \cdot \left[1 - f(t)\right] \tag{2b}$$

Here, J_{ms}^{para}, J_{sm}^{para} and, J_{ms}^{active}, J_{sm}^{active} are steady state unidirectional paracellular and cellular fluxes, respectively[4]. The time dependent functions, $g(t)$ and $f(t)$, are governed by the following initial and boundary conditions. For $t = 0$, $g(t) = f(t) = 1$, and for $t \rightarrow \infty$, $g(t) = f(t) = 0$. If $g(t)$ approaches zero faster than $f(t)$, separation of the fluxes flowing through the two pathways can be done by the following numerical method. For t_1 sufficiently small, during time interval, $0 < t < t_1$, and $\tau_1 < t_1$, $[\tau_1, M(\tau_1)]$ obeys:

$$M_{ms}(\tau_1) = J_{ms}^{para} \cdot [1 - g(\tau_1)] \qquad (3a)$$

$$M_{sm}(\tau_1) = J_{sm}^{para} \cdot [1 - g(\tau_1)] \qquad (3b)$$

where $\tau_1 < t_1$.

Similarly, for t_2 sufficiently large and $\tau_2 > t_2$, $[\tau_2, M(\tau_2)]$ obeys:

$$M_{ms}(\tau_2) = J_{ms}^{para} + J_{ms}^{active} \cdot [1 - f(\tau_2)] \qquad (4a)$$

$$M_{sm}(\tau_2) = J_{sm}^{para} + J_{sm}^{active} \cdot [1 - f(\tau_2)] \qquad (4b)$$

where $\tau_2 > t_2$.

Thus:

$$\frac{M_{ms}(\tau_1)}{M_{sm}(\tau_1)} = \frac{J_{ms}^{para}}{J_{sm}^{para}}, \text{ for } \tau_1 < t_1 \qquad (5a)$$

$$\frac{M_{ms}(\tau_2) - J_{ms}^{para}}{M_{sm}(\tau_2) - J_{sm}^{para}} = \frac{J_{ms}^{active}}{J_{sm}^{active}}, \text{ for } \tau_2 > t_2 \qquad (5b)$$

From definitions:

$$M_{ms}^{\infty} = J_{ms}^{para} + J_{ms}^{active} \qquad (5c)$$

$$M_{sm}^{\infty} = J_{sm}^{para} + J_{sm}^{active} \qquad (5d)$$

---

[4] The two cellular fluxes are given with superscript 'active'. This notation will be explained in *Section 3.3.2*.

The four *Eqs.* (5) allow for the four unknowns, J_{ms}^{para}, J_{ms}^{active}, J_{sm}^{para}, and J_{sm}^{active}, to be calculated.

3.1.2
Paracellular Convection

Nedergaard et al. (1999) used the strategy outlined above for studying the movement of radioactive alkali metal ions, $^{134}Cs^+$, ^{42}K, and $^{24}Na^+$, through isolated toad small intestine exposed to similar glucose containing Ringer's solution in luminal perfusion solution and serosal bath, see Fig. 2. The time course of unidirectional fluxes of $^{134}Cs^+$ could be fitted with monoexponential functions of comparable time constants, and the flux-ratio was time-invariant throughout the observation period, *i.e.*, from the first appearance of the isotope on the *trans* side to steady state isotope fluxes were achieved (upper left and right hand panels of Fig. 2). This result indicates that just the paracellular route is available for transport of $^{134}Cs^+$. With a similar protocol

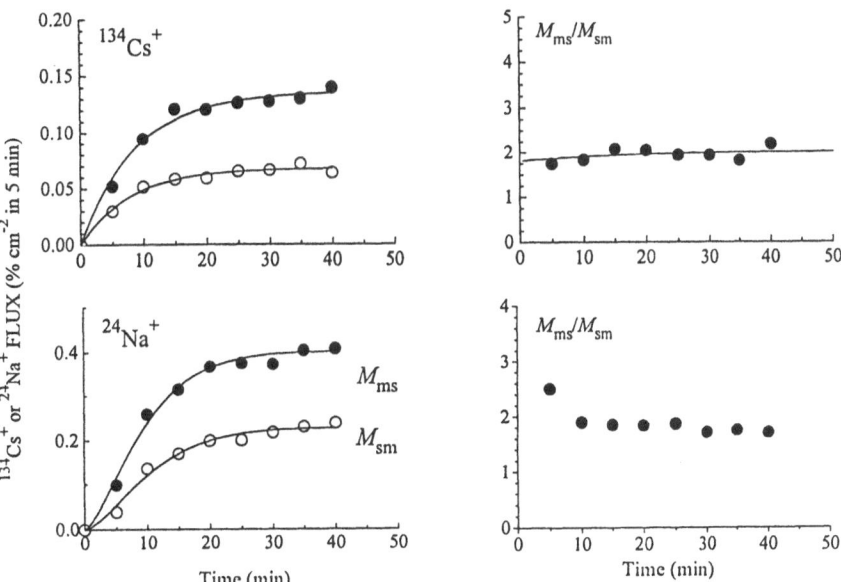

Fig. 2. Pre-steady state fluxes in isolated perfused toad (*Bufo bufo*) small intestine. Ringer's solution with 10 mM glucose on both sides (Nedergaard et al. 1999). *Upper left hand panel.* Appearing fluxes of $^{134}Cs^+$ added to mucosa (ms) or serosa (sm). *Upper right hand panel.* The ratio of appearing fluxes is time-invariant compatible with a single transepithelelial $^{134}Cs^+$ pathway. *Lower left hand panel.* Appearing fluxes of $^{24}Na^+$ added to mucosa (ms) or serosa (sm). *Lower right hand panel.* The ratio of appearing fluxes is time variant indicating more than one pathway for transepithelial Na^+ movement (from Nedergaard et al. 1999)

also pre-steady state fluxes of $^{42}K^+$ with dominating cellular pathways were studied. The tracer fluxes of this alkali metal ion never came to steady state within the observation period of 80 min – further strengthening the conclusion that $^{134}Cs^+$ does not trace the transcellular K^+ pathway. These studies of unidirectional $^{134}Cs^+$ fluxes also showed that there is a significant inwardly directed net flux of this ion in spite of the fact that a small transepithelial potential difference of, $\psi^m - \psi^s = -3.5 \pm 1.8$ mV [$N = 14$] would drive a net flux in outward direction. In the example of Fig. 2 (upper panels), the paracellular flux ratio was 2.0, and the average for 6 similar type of experiments was, $\left(J_{ms}^{para} / J_{sm}^{para} \right)_{Cs} = 2.06 \pm 0.12$ (mean $\pm$ SEM, $N = 6$). This number allows for the calculation of the driving force governing the net flux of $^{134}Cs^+$ through the paracellular pathway,

$$ W = \frac{R \cdot T}{F} \log_e \left(\frac{J_{ms}^{para}}{J_{sm}^{para}} \right) \tag{6} $$

The above flux ratio leads to, $W = 18.2 \pm 1.5$ mV. Considering the small transepithelial electrical potential difference of opposite sign (above), and the fact that the preparations were bathed with similar Ringer's solution on the two sides, it is indicated that transport of the cesium ion through the paracellular pathway takes place by convection.

3.1.3
Separation of Paracellular and Cellular Fluxes of the Sodium Ion

A similar type of experiments was performed with $^{24}Na^+$, and the example depicted in lower panels of Fig. 2 shows that unlike the ratio of unidirectional $^{134}Cs^+$ fluxes the ratio of unidirectional $^{24}Na^+$ fluxes is time-variant with a significantly larger flux-ratio initially (2.59) than at steady state (1.77). With the numerical method indicated by *Eqs.* (5) we estimated the four unidirectional fluxes (Table 1), and obtained an average ratio of paracellular unidirectional $^{24}Na^+$ fluxes of, $\left(J_{ms}^{para} / J_{sm}^{para} \right)_{Na} = 3.66 \pm 0.35$ ($N = 5$). This result confirms the conclusion derived from the cesium ion experiments, that there is a significant inward convective flow of small cations along the paracellular route. In conclusion, these experiments indicate that under transepithelial osmotic equilibrium there is an inward flow of water, which by convection moves tight junction-permeable cations in the same direction.

Table 1. Unidirectional Na$^+$ fluxes (pmol·s^{-1}·cm^{-2}) obtained by a method that does not perturb the epithelium's steady state (see Fig. 2). Glucose stimulated toad small intestine exposed to isotonic Ringer on the mucosal and the serosal side ($N = 5$ preparations, from Nedergaard et al. 1999)

	M_{ms}^{∞}	M_{sm}^{∞}	$\dfrac{M_{ms}^{\infty}}{M_{sm}^{\infty}}$	J_{ms}^{para}	J_{sm}^{para}	$\dfrac{J_{ms}^{para}}{J_{sm}^{para}}$	J_{ms}^{active}	J_{sm}^{active}
Mean	1330	560	2.57	450	130	3.66	880	430
± SEM	±120	± 100	0.26	± 40	± 20	±0.34	±130	± 90

* The steady state unidirectional fluxes through cellular and paracellular pathways were estimated by pre-steady state flux ratio analysis as explained in Chapter 3.1.1 [*Eqs.* (5)].

3.1.4
Paracellular Uphill Water Transport

The cesium flux-ratio method was used also for studying paracellular uphill water flow (Nedergaard 1999). In the experiments shown in Fig. 3 the osmotic pressure was raised on the mucosal side by adding urea (or mannitol, see below) at increasing concentrations (Figs. 3B and C). This resulted in a decrease of the flux-ratio (Fig. 3D), however, revealing that there is a large range of excess luminal osmotic concentrations where $\left(J_{ms}^{para} / J_{sm}^{para}\right)_{Cs} > 1$, and thus with the cesium ions submitted to inwardly directed convection. It is indicated, therefore, that the intestinal epithelium has capacity of generating paracellular convective flows of $^{134}Cs^+$ both under equilibrium conditions (isotonic transport) and under conditions of mucosal hypertonicity (uphill water transport). In another series of experiments 100 mM mannitol was added to the luminal bath (×-symbol of Fig. 3D) resulting in a flux ratio of 1.47 ± 0.15 ($N = 4$) as compared to 1.49 ± 0.21 ($N = 12$) with urea as mucosal osmolyte.

Thus, water transport in absorptive direction can take place even in the presence of an adverse osmotic gradient across the intestinal wall. This conclusion concerning toad small intestine confirms results of an early study by Parsons and Wingate (1958), who demonstrated that rat jejunum is capable of absorbing water from a luminal solution made hypertonic by varying its NaCl concentration. Furthermore, they showed that the flux of water and the luminal osmotic concentration were inversely related (as expected), and that

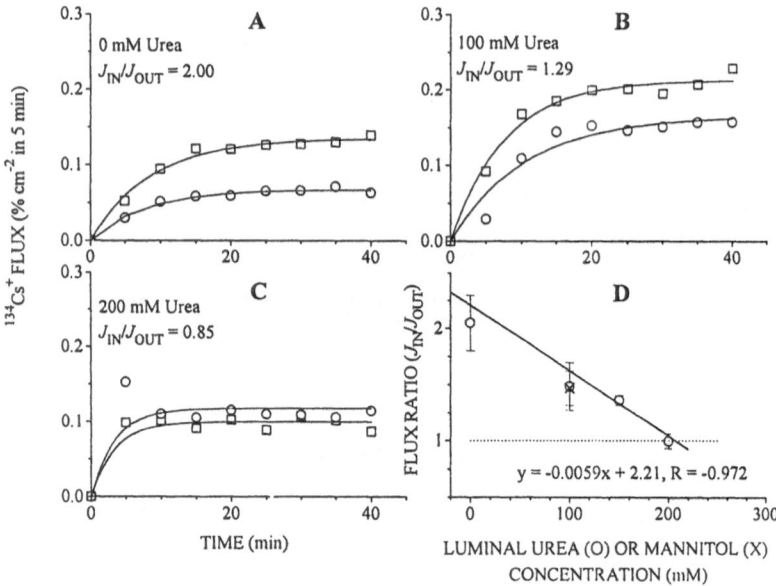

Fig. 3. Flux ratio analysis of convective Cs$^+$ transport across isolated perfused toad small intestine. Amphibian Ringer's solution on both sides with 10 mM glucose. **A:** With no osmotic difference across the preparations, a flux-ratio of 2.00 was obtained. **B:** With 100 mM urea added to the mucosal bath the ratio dropped, in this example to 1.29. **C:** With 200 mM urea added to the outside solution, the net flux virtually stopped, or reversed, the flux ratio was 0.85. **D:** Relationship between flux-ratio and urea concentration of luminal perfusion solution. The following flux ratios enter the analysis (mean ± sem). 0 mM Urea: 2.05 ± 0.14 (N = 6); 100 mM Urea: 1.49 ± 0.21 (N = 12); 150 mM Urea: 1.36 ± 0.43 (N = 3). 200 mM Urea: 0.944 ± 0.064 (N = 5). Mannitol: 1.47 ± 0.15 (N = 4). Equation of regression line: y = -0.00585 (± 0.00099)x + 2.21 (± 0.16). (From Nedergaard 1999)

water uptake was still significant with the (initial) lumen concentration being 130 mOsm above that of the (initial) serosal bath.

3.2
Other Studies on Vertebrate Small Intestine

3.2.1
Paracellular Transport

The conclusion above that a paracellular pathway is available for sodium ion transport is in agreement with previous studies of the rabbit ileum by Frizzell and Schultz (1972). They examined the voltage dependence of transepithelial unidirectional fluxes and found that large components pass by way of electrodiffusion through a single rate-limiting barrier, which were attributed to the paracellular pathway. Like in our studies of toad intestine (Neder-gaard et al. 1999) also in rat ileum is the paracellular flux of sodium ions smaller than the cellular Na^+ flux. A microelectrode study of flounder small intestine concluded that the paracellular pathway accounts for the major part of the transepithelial electrical conductance (Halm et al. 1985a). This need not be in conflict with the above conclusions of isotope studies that the major fraction of unidirectional Na^+ fluxes passes the cells. As will be dis-cussed below, the cellular pathway contains a number of different Na^+ transport systems that operate in an electroneutral fashion. Their contribu-tions to ion fluxes will be observed in isotope flux studies, only, not in elec-trical conductance measurements. More recent investigations (reviewed in Pappenheimer 1993) have indicated that the tight-junction permeability to hydrophilic solutes is regulated. Specifically, the permeability increases when glucose and amino acids are present in the luminal solution. In the glucose-activated state of ileal mucosa and under transepithelial thermody-namic equilibrium conditions monosaccharides and amino acids are also moving inward between the cells by means of convection (Pappenheimer and Reiss 1987, Pappenheimer 1987) with the lateral intercellular space dramatically expanded (Atisook et al. 1990). Pappenheimer and coworkers' general conclusion, that convection governs fluxes of tight junction-permeable ions and non-electrolytes, is compatible with our conclusion of paracellular convection fluxes of $^{134}Cs^+$ and $^{24}Na^+$.

The transport of sodium from the intestinal epithelial cells into the sero-sal bath is by way of primary active transport via the ouabain inhibitable Na^+/K^+-pump (Schultz and Zalusky 1964, Fordtran and Dietschy 1966, Schultz and Curran 1968, Stirling 1972, Wright et al. 1979). Using rings or mucosal sheets of rabbit small intestine dissected free of the heavy muscle layers and ^3H-ouabain of high specific activity, Stirling (1972) provided the evidence that radiolabelled ouabain binds specifically to the membranes facing the lateral intercellular space, and not to the brush border membrane. This is in contrast to the phlorizin-^3H binding sites that with a similar tech-nique were localized exclusively on the brush border membrane (Stirling 1967, Stirling et al. 1972). It follows that radiolabelled sodium ions having entered the cells from the mucosal solution, are being pumped into the lat-eral space before they are transported further into the serosal bath by con-vection-diffusion (*Section* 2.3.1).

In a study of frog skin epithelium it was found that the amount $^{134}Cs^+$ trapped by the epithelial cells is proportional with the activity of the sodium pump indicating that (*i*) radiolabelled cesium ions are taken up by cells via Na^+/K^+-pumps and (*ii*) once $^{134}Cs^+$ has entered the cells this ion is literally irreversible trapped by the cells (Nedergaard et al. 1999). Thus, in toad intestine it is to be expected that a fraction of $^{134}Cs^+$ entering the lateral space from the luminal bath via tight junctions is being pumped into the cells rather than proceeding downstream to the serosal bath. In our experimental study it was found – in agreement with Na^+/K^+-pumps being expressed on the membranes lining the lateral intercellular space – that $^{134}Cs^+$ added to the mucosal bath indeed are being trapped by cells. It was also found that the amount of $^{134}Cs^+$ trapped by cells is increasing linearly with the inward convective flow of this tracer. This finding would be compatible with the notion that the activity of these pumps governs the rate of water flow across the epithelium (Nedergaard et al. 1999).

3.2.2
Cellular Transport Systems

3.2.2.1
The Luminal Membrane

Both Na^+ and Cl^- as well as glucose and amino acids are transported across the epithelial cells by metabolic energy requiring mechanisms. Apical transport of Na^+-gradient coupled metabolites, monosaccharides and amino acids, have been studied with respect to selectivity, maximum transport capacity, and stoichiometry; these systems are highly specific for Na^+ (the early literature is reviewed in Schultz and Curran 1970, 1974). The Na^+-dependent uptake of neutral amino acids and sugars results in a depolarization of the apical membrane indicating that the transport of these substances is associated with the movement of positive charges across the membrane, *i.e.*, the Na^+-coupled transport of the hydrophilic nonelectrolytes is rheogenic (Rose and Schultz 1971, White and Armstrong 1971, Schultz 1977, Okada et al. 1977, Gunter-Smith et al. 1982). The gene coding for the apical Na^+-glucose transporter has been cloned (Hediger et al. 1987), and studies in Wright's laboratory of the gene product, SGLT1, in the *Xenopus laevis* oocyte expression system have resulted in a model for Na^+-glucose transport (Parent et al. 1992a,b, Wright 1993). Using GLUT-expressing J774 murine macrophages and SLGT1-expressing *Xenopus* oocytes, Fischbarg and coworkers provided the evidence that both the facilitated glucose transporter (GLUT) and SLGT1 behave as water channels (Fischbarg et al. 1989, 1990, 1993). Further to these studies, Loo et al. (1996) found that in a hypotonic solution the volume of SLGT1-expressing *Xenopus* oocytes increases when the Na^+-glucose transporter is being activated. These and the more recent studies of water transport associated with activation of the

SLGT1-protein are reviewed in another paper of the present review-series (Zeuthen 2000) and, therefore, they shall not be discussed further in our review.

The apical uptake of Cl⁻ takes place against this ion's electrochemical potential gradient and as an electroneutral process coupled to the entry of Na^+ (Nellans et al. 1973, Frizzell et al. 1979). In intestinal epithelia of several vertebrate species the electroneutral Na^+ and Cl⁻ entry into the cells from the luminal fluid involves a set of two counter transporters, a Na^+/H^+- and a Cl⁻/HCO_3^- exchanger, respectively (Fordtran et al. 1968, Turnberg et al. 1970, Liedtke and Hopfer 1982b, Knickelbein et al. 1985). At molecular level the apical Cl⁻/HCO_3^- exchanger has not yet been characterized, but the gene coding for the Na^+/H^+ exchanger has been cloned. The gene product, NHE3, is different from two other isoforms of this family, NHE1 and NHE2, expressed in non-polar cells and in the inner membrane of epithelial cells (Sardet et al. 1989, Orlowski et al. 1992, Tse et al. 1991, Tse et al. 1992). The notion of two parallel counter transporters, rather than a single cotrans-porter, is compatible with the observation (Frizzell et al. 1973) that the car-bonic anhydrase inhibitor, acetazolamide, inhibits the active entrance of Cl⁻ in rabbit ileum *in vitro*. Probably, the coupling of the Na^+ and Cl⁻ uptake by these two mechanisms is mediated by the carbonic anhydrase catalyzed hydration of CO_2 which would provide the two cellular substrates, H^+ and HCO_3^-, for the apical Na^+/H^+- and Cl⁻/HCO_3^- exchangers, respectively (Knickelbein et al. 1985). A K^+-independent thiazide-sensitive Na-Cl co-transporter has been isolated from flounder urinary bladder and a similar transporter seems functioning in rat kidney (Gamba et al. 1993). Functional studies indicated that this transporter is not present in mammalian small intestine (Liedtke and Hopfer 1982a). In flounder and *Amphiuma* small intestine, however, a major fraction of the ³⁶Cl⁻ uptake was dependent on mucosal Na^+ and K^+ (requiring both cations) and it was inhibited by fu-rosemide, indicating that a Na^+-K^+-Cl⁻ cotransporter is operating in parallel with the apical anion exchanger and apical K^+ channels (Musch et al. 1982, Halm et al. 1985b, White 1989). The stoichiometry of the cotransport is $1Na^+:1K^+:2Cl^-$ and it is also bumetanide sensitive (O'Grady et al. 1986, O'Grady et al. 1987). Purified from the winter flounder intestinal brush-border membrane it is a 170–175 kDa protein, which in contrast to other co-transporters, is deactivated by *cyclic*GMP-dependent phosphorylation (Suvitayat et al. 1994). Thus, in vertebrate small intestine there seems to be different solutions to the electroneutral uptake of Na^+ and Cl⁻. Further to this, it should be mentioned that mammalian intestine exhibits capacity for HCO_3^- secretion (ileum) as well as HCO_3^- absorption (jejunum). For studies and reviews on HCO_3^- transport and this ion's regulatory role for other transporters we refer to White et al. 1984, Hopfer and Liedtke 1987, and Sullivan and Field 1991.

The cellular mechanism of intestinal Cl⁻ secretion has been studied in short-circuited mammalian small intestine with experimentally elevated cellular [cyclicAMP] accomplished by stimulation of the adenylate cyclase, inhibition with theophylline of cyclicAMP-breakdown, or by adding a membrane permeable cyclicAMP analogue to the serosal bath. The early studies (Field et al. 1968, Field 1971, Field 1974, Sheerin and Field 1975, Frizzell et al. 1979) showed that these protocols may enhance the transepithelial secretory ion flows to such an extend that secretion, not absorption, of electrolytes becomes the dominating function. The secretory flux requires Na^+ in the serosal bath and is inhibited by furosemide, or ouabain. Also elevation of the concentration of intracellular free-Ca^{2+} by the Ca^{2+} ionophore A23187 was shown to stimulate Cl⁻secretion. Studies of rat (Liedtke and Hopfer 1982a) and pig (Forsyth and Gabriel 1989) brush-border vesicle preparations, and intracellular microelectrode studies of small intestine of urodeles (*Amphiuma*, *Necturus*, White 1986, 1994, Giraldez et al. 1988) identified a Cl⁻ conductance of the brush border membrane, and Giraldez et al. 1988 showed that this conductance could be activated by application of forskolin or the membrane permeable dibutyryl cyclicAMP to the serosal bath. In the same laboratory single chloride selective channels were studied in cell-attached and excised inside-out membrane patches of vertebrate enterocytes (Giraldez et al. 1989). Currents were recorded from outward rectifying Cl⁻ channels with a conductance of 17–25 pS, which were activated by raising cellular [cyclicAMP]. The channel was not gated by cytosolic Ca^{2+} or by membrane voltage. However, chloride channels in excised patches could also be activated by large membrane depolarizations. With cellular [Cl⁻] being above equilibrium the Cl⁻ selective channels would constitute an exit pathway for the secretory Cl⁻ flux. Studies on intestinal chloride channels, including the phosphorylation activated cystic fibrosis transmembrane conductance regulator (CFTR), have been reviewed by Grubb and Boucher (1999). The major information on secretory anion channels, however, has been obtained with other secretory epithelia or cultured cells expressing different types of secretory chloride channels (reviews of these studies are found in Gögelein 1988, Frizzell and Morris 1994, Dawson et al. 1999, Gadsby and Nairn 1999, Schwiebert et al. 1999, and Sheppard and Welsh 1999).

3.2.2.2
The Inner Membrane

Glucose and amino acids, which by cotransport with Na^+ in the brush border membrane are accumulated above equilibrium in the cells, are presumably translocated downhill to the serosal side by specific transport systems ('facilitated diffusion'). The glucose transporter of the inner membrane, GLUT2, has been cloned. It is a member of a large family of facilitated glucose transporters of which the first, GLUT1, was cloned from human erythrocytes, and GLUT5 is a fructose transporter located in the apical membrane (reviewed by Thorens 1993). Members of this family of transporters operate without Na^+, so that the exit of glucose from the intestinal cells to the serosal bath is not coupled to a transport of Na^+ or any other ion.

Since the first studies compiled and discussed in Frizzell et al. (1979), it has been assumed that Cl^- and K^+ both move by electrodiffusion from the cells into the serosal bath. Studies by Halm et al. (1985b), however, have indicated that exit of Cl^- and K^+ in flounder intestine may take place as electroneutral K-Cl co-transport. Intracellular recordings combined with ion substitution protocols indicated a Cl^- conductance of the serosal membrane in flounder (Halm et al. 1985a), but not in *Amphiuma* (White and Ellingsen 1989) intestinal cells. The last mentioned study revealed a significant K^+ conductance of this membrane, which apparently was absent in flounder intestinal cells (Halm et al. 1985b). In a patch clamp study of goby intestinal epithelial cells dissociated by an enzymatic method currents were recorded from depolarization activated anion selective channels (Loretz and Fourtner 1988). Taken together, the above studies indicate that the regulation of the inner membrane's conductive properties and ion selectivity is, as yet, not very well understood, perhaps because intestinal epithelia from different vertebrate classes, or species, do not express similar ion channels and other transporters.

In animal cells, several transporters are responsible for coupled transport into the cells of Na^+, Cl^-, and K^+. The intestinal cotransporter, NKCC1, is an isoform that possibly is found more generally, perhaps exclusively, in the serosal membrane of polarized cells in secretory and absorptive epithelia. The co-transporter in shark rectal gland (Hannafin et al. 1983, Greger and Schlatter 1984ab) was the first to be cloned, and it was more recently identified in the mouse inner medullary collecting duct cell line (mIMCD-3) and in T84-cells, which are not derived from small intestine but constitute a human colonic carcinoma cell line. The two mammalian transporters share 71-74% amino acid sequence identity with that of the shark (reviewed in Haas 1993, Payne and Forbush 1995).

3.2.3.
Cellular Heterogeneity

From the studies reviewed above it is clear that vertebrate small intestine is configured both for Cl⁻ absorption and Cl⁻ secretion. In reviews by Madara (1991) and Sullivan and Field (1991), and in analogy with what has been suggested for the epithelium of large intestine (Welsh et al. 1982), it was assumed that in the mammalian small intestine NaCl-absorption and uptake of nutrients are located to the surface cells whereas the secretory capacity is confined to the proliferating ('undifferentiated') crypt cells. This notion of functional heterogeneity among intestinal cells is supported by the finding that *in situ* hybridization in rat small intestine with ^{35}S-radiolabelled probes of CFTR mRNA synthesized from rat cDNA (Trezise and Buchwall, 1991) and in human small intestine with ^{35}S-mRNA probes synthesized from *h*CFTR cDNA (Strong et al. 1994) indicated in both species a decreasing expression of CFTR from the crypt cells to the surface cells. Matthews et al. (1998) found that the messenger of the Na^+-K^+-$2Cl^-$ cotransporter (NKCC1) was abundantly expressed in rat jejunal crypt cells, but undetectable in cells from the tip of the villi. By targeting mRNA and not functionally expressed membrane proteins, these studies do not rule out unequivocally, however, that the surface cells exhibit some Cl⁻ secretion. Electrophysiological studies indicated quantitative, but not qualitative differences of conductance expression along colonic crypt axis of rat with increasing amiloride-sensitive Na^+ conductance and decreasing cyclicAMP activated Cl⁻ conductance from base to surface resulting in a limited ability of the surface cells to secrete Cl⁻ (Ecke at al 1996, Greger et al. 1997). An intracellular microelectrode study of *Necturus* small intestine reported a significant membrane depolarization of cells in which an apical Cl⁻ conductance was stimulated simultaneously with the stimulation of alanine uptake (Giraldez and Sepulveda 1987), indicating that absorption and secretion also in this species are co-expressed. Furthermore, patch clamp studies have disclosed chloride channels in the mucosal membrane of the rat colon epithelium (Diener et al. 1989), and with the self-referencing probe technique for localizing extracellular ion currents in hen coprodeum it was concluded that in this intestinal epithelium, active Na^+ absorption and Cl⁻ secretion occur from the same area/cells (Holtug et al. 1991). Although molecular biologic studies indicate that expression of membrane proteins involved in ion secretion by intestinal cells of vertebrate gastrointestinal tract is being depressed at gene level during migration of the cells along the crypt-villus axis, studies of secretory and absorbtive functions along this axis indicate that secretory activity can still be evoked in cells with predominant absorptive function. Accordingly, the cell model of the epithelium of toad small intestine to be discussed next contains membrane pathways expressing both functions.

3.3
Interpretations of Na⁺ Fluxes Across Small Intestine

3.3.1
The Transport Model

The cellular model of vertebrate intestinal mucosa shown in Fig. 4 compiles results of studies discussed above, but we hypothesize that transport proteins are expressed differently in the lateral and the serosal membranes, with dominant expression of Na^+/K^+-pumps on the lateral membrane and dominant expression of Na^+-K^+-$2Cl^-$cotransporters on the serosal membrane. The first step of NaCl absorption is the electroneutral uptake across the apical membrane. Assuming expression of Cl^- channels on the lateral mem-

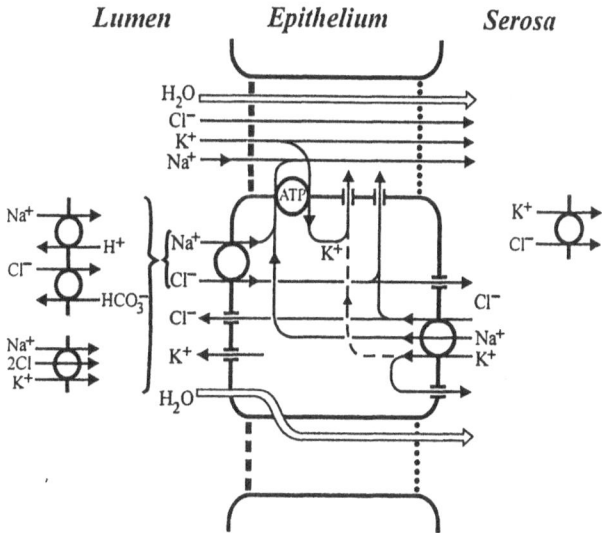

Fig. 4. Ion flux pathways in vertebrate small intestine, including recirculation fluxes associated with transepithelial (paracellular) fluid transport. In this model of small intestine water flows between the cells while the ion fluxes contains both paracellular convective components, and cellular active components. *Inset left*: The electroneutral uptake of NaCl is mediated by either of two mechanisms: a set of double exchangers, or as co-transport with K^+. *Inset right*: An electroneutral exit of KCl has been proposed for the inner membrane (see *Sections 3* for further details). From Nedergaard et al. 1999

brane the second step is the pumping of Na^+ through the lateral membrane and the downhill transport of Cl^- from cells into the lateral space. Finally, the two ions are transported by convection-diffusion into the serosal bath. Due to the activity of a cotransporter in the serosal membrane, at steady state the cellular fluxes of the sodium ion are both inwardly and outwardly directed. Both fluxes pass lateral pumps and are therefore active. With expression of pumps on lateral membranes it is likely that also K^+ channels are located on this membrane domain. Owing to the substrate requirement of the cotransporter in the serosal membrane, sodium ions, as well as potassium ions and chloride, are recirculated across the inner border of the epithelium.

By pump-dependent building up of a [NaCl] of lateral space, above that of bath, the condition for paracellular water transport is established. Unlike other epithelia, immunocytochemical studies have not located water channels in small intestine (Nielsen et al. 1993). A recent theoretical analysis showed that the demonstrated paracellular convective flows of $^{134}Cs^+$ and ^{24}Na would be compatible both with transjunctional and translateral transport of water (Larsen et al. 2000). In the model of Fig 4 we have kept the possibility open that water may also flow through cells. What is of more importance here is that in the absence of transepithelial concentration, electric, and hydrostatic gradients the paracellular water flow provides the condition for convection-diffusion fluxes of tight junction-permeating ions and hydrophilic organic molecules from the lateral space into the serosal solution. In a later section it will be discussed that this model also has capacity of transporting water in the inward direction when exposed to a hypertonic luminal fluid.

In the following section, first we derive equations for calculating the recirculation flux of Na^+. The results will then be used for estimating the virtual tonicity of the fluid emerging downstream the lateral space.

3.3.2
Calculation of the Na^+ Recirculation Flux

With reference to the model of Fig. 4, the inward active Na^+ flux, J_{ms}^{active}, is entering the cell through the apical membrane and is pumped by the Na^+/K^+ pump into the lateral intercellular space from where it proceeds to the serosal bath. This flux is depicted in Fig. 5, *upper part* together with the "invisible" flux, $J_{ms}^{active, return}$, returning via the Na^+ permeable junction membrane to the mucosal solution. The outwardly directed active Na^+ flux is transported into the cell from the serosal bath by the Na^+-K^+-$2Cl^-$ cotransporter and is pumped from the cell further into the lateral space by the Na^+/K^+-pump for proceeding to the mucosal bath via the junction membrane. This flux, J_{sm}^{active}, is depicted in Fig. 5, *lower part* together with the "invisible" flux, $J_{sm}^{active, return}$, returning to the serosal bath through the barrier downstream the lateral intercellular space. Observing that lateral pumps represent a source that delivers an isotope at constant rate in to a reversible pathway, we can make use of the theoretical result (Ussing 1952) that this does not affect the ratio of fluxes appearing at the two ends of the pathway. Thus, the active and "invisible" fluxes together with the appearing fluxes obey the following relationships:

$$\frac{J_{ms}^{active}}{J_{ms}^{active, return}} = \frac{J_{ms}^{para}}{J_{sm}^{para}} \qquad (7a)$$

$$\frac{J_{sm}^{active, return}}{J_{sm}^{active}} = \frac{J_{ms}^{para}}{J_{sm}^{para}} \qquad (7b)$$

With cellular and paracellular fluxes taken from Table 1, *Eqs.* (7) were used to calculate the "invisible" (return) flux components, see Table 2. The pumped flux is given by (Fig. 5):

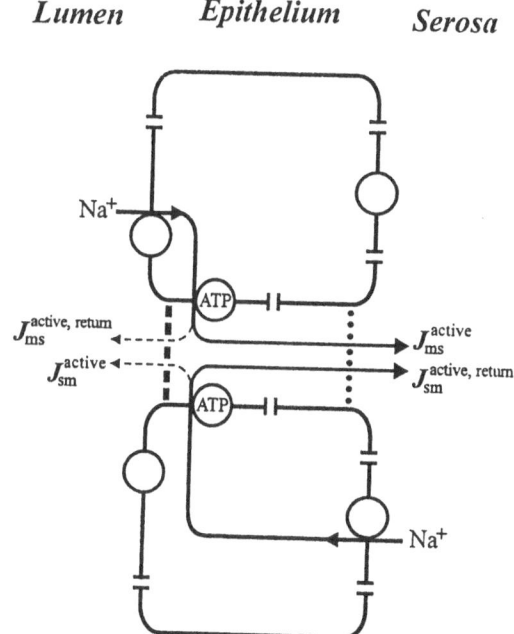

Lumen   *Epithelium*   *Serosa*

Fig. 5. Analysis of forward and backward cellular pathways of unidirectional $^{24}Na^+$ fluxes. *Upper part*: Influx path. *Lower part.* Outflux path. J_{ms}^{active} and J_{sm}^{active} are the forward and backward active fluxes, respectively. The 'return' fluxes ($J_{ms}^{active, return}$ and $J_{sm}^{active, return}$) are 'invisible' components, which can be calculated by *Eqs.* (10). From Nedergaard et al. (1999)

Table 2. Sodium ion recirculation in unperturbed toad small intestine. Listed are the "invisible" return fluxes according to definitions given in Fig. 5, together with the fraction of the pump flux, which is derived from the serosal bath [conf. text *Eq.* (11)]. Fluxes listed in column 2 and 3 are in $pmol \cdot s^{-1} \cdot cm^{-2}$ (From Nedergaard et al. 1999)

	$J_{sm}^{active, return}$	$J_{ms}^{active, return}$	$\dfrac{J_{Na}^{Return}}{J_{Na}^{Pump}}$
Mean	1730	260	0.65
±SEM	±320	± 60	±0.02

$$J_{Na}^{Pump} = J_{ms}^{active} + J_{ms}^{active, return} + J_{sm}^{active} + J_{sm}^{active, return} \quad \text{with}$$

$$J_{Na}^{Return} = J_{sm}^{active} + J_{sm}^{active, return}$$ coming from the serosal bath. Accordingly, the fraction of the pumped flux that is coming from the serosal bath is given by:

$$\frac{J_{Na}^{Return}}{J_{Na}^{Pump}} = \frac{J_{sm}^{active} + J_{sm}^{active, return}}{J_{ms}^{active} + J_{ms}^{active, return} + J_{sm}^{active} + J_{sm}^{active, return}} \tag{8}$$

According to this interpretation of estimated fluxes, on the average, 65% of the sodium ions pumped through the lateral membrane is being derived from the serosal solution (Table 2). This component represents the Na^+ recirculation-flux.

3.3.3
Virtual Osmotic Concentration of the Fluid Emerging from Lateral Space

In response to manipulating the transepithelial gradients of osmotic pressure Curran and Solomon (1957) found that water is moving freely across the wall of perfused intestinal epithelium of anesthetized rats *in vivo*, that absorption of water is associated with absorption of NaCl, and that the rates of water and solute absorption result in an isotonic concentration of the absorbate. In subsequent *in vitro* studies Curran (1960) further showed that fluid is flowing in the inward direction in absence of an external osmotic gradient, but if the net solute uptake was stopped, net water movement also stopped. He concluded that indirectly by depending on the active uptake of salt, is also the uptake of fluid depending on energy metabolism of the intestinal epithelial cells. This major principle of the Curran model (Fig. 1B) is contained the model depicted in Fig. 4. Thus, it is by maintaining a hypertonic lateral intercellular fluid the sodium pump is driving water into the lateral space, with the subsequent transport of fluid from the lateral space into the serosal bath forced by the excess hydrostatic pressure of the lateral space. The methods used in our study have resulted in an estimate of the flux pumped into the lateral space. The astonishing conclusion is that a significant fraction, according to our estimate about 65% of the pumped sodium ions is derived from the serosal bath. We will now ask the question: What would be the osmotic concentration of the fluid emerging from the downstream lateral space, assuming isotonic transport (Curran 1960)?

Without affecting the general outcome of the calculations below, we will simplify the treatment by disregarding in extracellular compartments all other ions than Na$^+$ and Cl$^-$. Thus, we can write for each of the three extracellular compartments, $C_{Na} = C_{Cl} \equiv C_{NaCl}$, with the nominal osmotic pressure of a given compartment being twice its concentration of NaCl. The flux of Na$^+$ added to the serosal bath, M^{Δ}, expressed as fraction of the flux, J_{Na}^{bm}, emerging from the lateral intercellular space, is given by,

$$\frac{M^{\Delta}}{J_{Na}^{bm}} = \frac{M_{ms} - M_{sm}}{(J_{ms}^{active} + J_{sm}^{active,\,return}) + (J_{ms}^{para} - J_{sm}^{para})} \tag{9a}$$

Assuming an isotonic absorbate (Curran 1960), it follows that $M^{\Delta} = J_v \cdot C_{Bath}^{Na}$, and:

$$\frac{J_{Na}^{bm} + J_{Cl}^{bm}}{J_v} = 2 \cdot C_{NaCl}^{bath} \frac{(J_{ms}^{active} + J_{sm}^{active,\,return}) + (J_{ms}^{para} - J_{sm}^{para})}{M_{ms} - M_{sm}} \tag{9b}$$

With the fluxes listed in Tables 1 and 2, and C_{NaCl}^{bath} = 120 mM corresponding to a nominal osmotic concentration of 240 mOsm, we calculate [Eq. (9b)] that the *virtual* osmotic concentration of the fluid leaving the lateral space would be, $(J_{Na}^{bm} + J_{Cl}^{bm})/J_v$ = 780 mOsm, *i.e.* 780–240 = 540 mOsm hypertonic with reference to bath. In the following *Section 4*, by a different approach we analyze this result and show that it is not violating the requirement that the osmotic concentration of the lateral intercellular space cannot – under any circumstances – be that much above the osmotic concentration of the bathing solutions.

4
Features of a Convection-Diffusion Process
with Implications for Solute Coupled Water Transport

4.1
The Convection-Diffusion Equation for a Non-Electrolyte
in a Homogenous Regime

Ion movements through tight junction and interspace basement membrane may be caused by diffusion, migration under the influence of an electric field, and convection. The purpose of this section is to briefly present and discuss simplifying equations that can be applied for dealing with this type of transport, and outline how they can be used for analyzing relationships between solute and water fluxes and intraepithelial solute and pressure gradients of a compartment model of the leaky epithelium that comprises electroneutral solutes[5]. Thus, in this version of the model effects of electric fields on solute fluxes are disregarded. Because the transepithelial electrical potential difference is close to zero and with electroneutral co- and counter transporters in the plasma membranes, modelling electroneutral transports is expected to provide general insight into the coupling of solute and water fluxes in the leaky epithelium. Obviously, this type of model cannot examine features related to electrical properties, which distinguish the leaky from the tight epithelium.

The equation for a stationary flux, J_S, in a homogenous membrane governed by diffusion with superimposed convection can be written (Patlak et al. 1963):

$$J_S = J_V (1-\sigma) \frac{C^{(1)} \cdot \exp[J_V (1-\sigma)/P_S] - C^{(2)}}{\exp[J_V (1-\sigma)/P_S] - 1} \qquad (10a)$$

Here P_S is the permeability coefficient of S in the membrane, $J_V \cdot (1 - \sigma)$ the convection velocity of solute with σ being the reflection coefficient ($0 \le \sigma < 1$), and $C^{(1)}$ and $C^{(2)}$ concentrations of the external compartments (1) and (2). Eq. (10a), which assumes transport in an electroneutral regime, was used by Patlak et al. (1963) in their analysis of coupling of solute and water flows in a compartment of the type presented in Fig. 1A. It was applied also in an extended mathematical model of convection-diffusion fluxes across tight junction and interspace basement membrane of an epithelium with recirculation of the driving species via adjacent cells (conf. Fig. 1C, and Lar-

---

[5] The presentation and conclusions build on the paper by Larsen et al. 2000.

sen et al. 2000). We will discuss the model below. But first it will be shown that the question raised immediately above about the relationship between concentration of the lateral space and the virtual concentration of the trans-portate emerging from the lateral space can be examined by analysis of Eq. (10a) and expressions derived from this equation.

The relationship given by Eq. (10a) between solute flux and volume flow for a convection-diffusion process across a single membrane has Fick's law as one of its limits, i.e., $lim\ J_S = P_S \cdot (C^{(1)} - C^{(2)})$, for $J_V \to 0$. In this limit, with no water flow, only the solute passes the membrane (the virtual concentra-tion of the transportate is infinitely large). In the other limit where the proc-ess is dominated by convection, i.e., $J_V \to \pm \infty$, it can be shown that, $lim\ J_S/J_V = C^{(1)} \cdot (1 - \sigma)$ and $lim\ J_S/J_V = C^{(2)} \cdot (1 - \sigma)$, respectively. Thus, with pure convection and a small value of σ (large pores), the concentration of the transportate flowing into the *trans* compartment is almost identical with the concentration of the *cis* compartment. In the real situation the concentration of the transportate must be somewhere between the above mentioned two extremes. With both unidirectional fluxes being positive, their ratio is given by (Larsen et al. 2000):

$$\frac{J_S^{1 \to 2}}{J_S^{2 \to 1}} = \frac{C^{(1)}}{C^{(2)}} \exp(J_V(1-\sigma)/P_S) \qquad (10b)$$

Thus the ratio of unidirectional fluxes governed by a convection-diffusion process is depending on the ratio of volume flow and permeability coeffi-cient. This relationship is illustrated in Fig. 6A showing the variation of the flux-ratio with the argument of its exponential term, $J_V \cdot (1 - \sigma)/P_S$. As the volume flow decreases relative to the permeability coefficient the flux-ratio also decreases. In Fig. 6B is shown the relationship between the resulting concentration of the transportate, J_S/J_V, and $J_V \cdot (1 - \sigma)/P_S$. As the argument, $J_V \cdot (1 - \sigma)/P_S$, attains values corresponding to flux-ratios in the order of e, that is, corresponding to values obtained in our studies of paracellular fluxes in toad small intestine, the virtual concentration of the transportate is be-coming significantly larger than the concentration of the solute in the com-partment from which the fluid is derived. Thus, with volume flow rates small compared to the diffusion coefficient of the solute is diffusion dominating over convection so that, $J_S/J_V \gg C^{(1)} \cdot (1 - \sigma)$. In the above flux-ratio regime, compartment (1) may represent the lateral intercellular space of the toad small intestine, and compartment (2) the serosal bath. Then with fixed re-flection coefficient, isotonicity of the transportate would be achieved if the solute is being recirculated via cells.

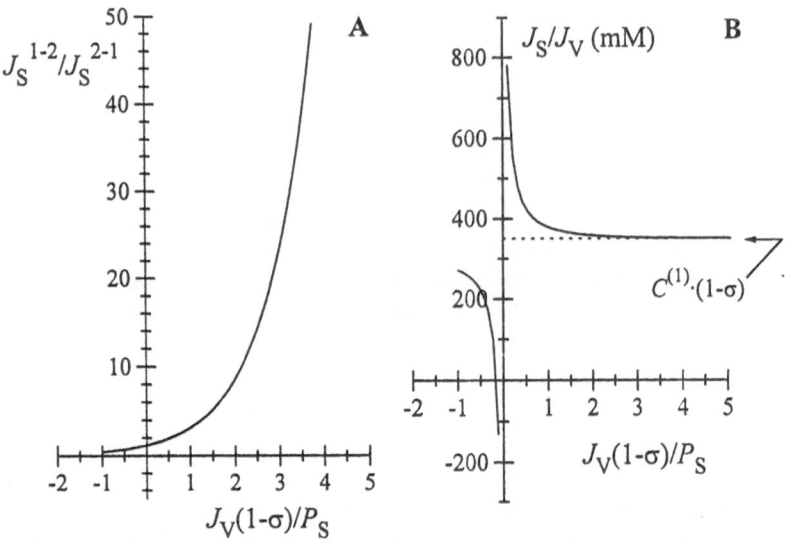

Fig. 6. *A*: Ratio of unidirectional fluxes of an electroneutral solute moving by convection-diffusion across a homogeneous membrane. The ratio is here depicted as function of the ratio between volume flow (times 1-σ) and diffusion coefficient divided by membrane thickness. With a small diffusion coefficient relative to volume flow, the flux-ratio attains very large values. *B*: With a small reflection coefficient (large pores) and in the limit of large volume flow relative to the permeability coefficient is the concentration of the emerging fluid close to that of the compartment from which the fluid is derived (compartment 1). In the regime corresponding to biological preparations, ($J_s^{1 \to 2} / J_s^{2 \to 1}$) ≈ *e*, resulting in a concentration of the emerging fluid that is significantly larger than that of the compartment from which the fluid is derived

For exploring how the principles outlined above affects solute coupled water transport in a leaky epithelium we will now discuss computations based on a mathematical model of the epithelium that incorporates convection-diffusion pores in the membranes delimiting the paracellular route and a recirculation pathway in the serosal membrane.

4.2
The Mathematical Description of a Compartment Model

The quantitative analysis was performed on the basis of the model of the leaky epithelium depicted in Fig. 7 (Larsen et al. 2000). It contains 4 well-stirred compartments: mucosal (*o*), serosal (*i*), cell (*c*), and lateral intercellular space (*lis*), which define 5 membranes: apical (*a*), serosal (*s*), lateral (*lm*), tight junction (*tm*), and the interspace basement membrane (*bm*). The non-charged driving solute (S) is transported into *lis* by a saturating pump conventionally described by:

$$J_S^{\text{pump}} = J_S^{\text{max,pump}} \left(\frac{C_S^c}{K_S^{\text{pump}} + C_S^c} \right)^3 \tag{11a}$$

where $J_S^{\text{max,pump}}$ is the pump flux with all sites occupied and K_S^{pump} is a dissociation constant. S is taken up by the cells by Fick diffusion through mucosal and serosal membrane:

$$J_S^m = P_S^m (C_S^{(1)} - C_S^{(2)}) \tag{11b}$$

In *Eq.* (11b), for diffusion across the apical membrane are $m = a$, (1) = *o* and (2) = *c*, and for diffusion across serosal membrane are m = s, (1) = *cell* and (2) = *i*. Fluxes from (1) to (2) are positive. In some calculations a pure diffusion channel with diffusion permeability, $P_S^{\text{tm,diff}}$, is also assumed in the tight junction membrane carrying a diffusion flux given by:

$$J_S^{\text{tm,diff}} = P_S^{\text{tm,diff}} \cdot (C_S^o - C_S^{\text{lis}}) \tag{11c}$$

With notations similar to those indicated above for *Eq.* (11b), the convection-diffusion flux through tight junction and interspace basement mem-

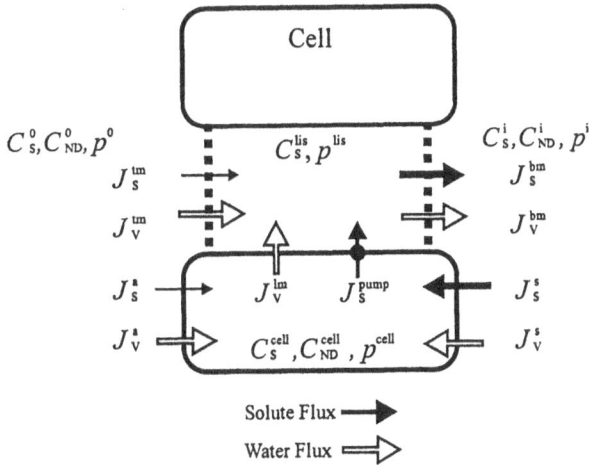

Fig. 7. Symbols used in the mathematical model of solute coupled fluid transport in an absorbing epithelium. Arrows indicate direction of positive solute and water fluxes. Equations of the fluxes across the four boundaries are given in the text. (From Larsen et al. 2000)

brane were calculated from *Eq.* (10a). Thus, with a pure diffusion pore [*Eq.* (11c)] as well as a convection-diffusion pore [*Eq.* (10a)] we can analyze the two cases of tight junction membranes, *i.e.*, solutes enter the lateral intercellular by diffusion, only (translateral water transport), or they enter this space by convection-diffusion (transjunctional water transport). The water flux through the five membranes *a*, *s*, *lm*, *tm*, and *bm*, are calculated from:

$$J_V^m = L^m \left[RT \left\{ \sigma^m (C_s^{(2)} - C_s^{(1)}) + (C_{ND}^{(2)} - C_{ND}^{(1)}) \right\} + (p^{(1)} - p^{(2)}) \right] \tag{12}$$

where C_{ND} is the concentration of a non-charged, non-diffusible solute in compartment indicated by superscript, and $\sigma^a = \sigma^s = \sigma^{lm} = 1$. In all computations similar hydrostatic pressure of outer and inner compartments were assumed[6]:

$$p^{cell} = p^o = p^i \tag{13}$$

Mass conservation of the diffusible solute and water, respectively, are expressed in 4 equations:

$$J_S^{pump} = J_S^a + J_S^s \tag{14a}$$

$$J_S^{bm} = J_S^{tm} + J_S^{pump} + J_S^{tm,\,diff} - J_S^{bm,\,diff} \tag{14b}$$

$$J_V^a = J_V^s + J_V^{lm} \tag{14c}$$

$$J_V^{bm} = J_V^{tm} + J_V^{lm} \tag{14d}$$

The tonicity, *TON*, of the net transportate is given by:

$$TON = \frac{J_S^{bm} + J_S^s}{J_V^{bm} + J_V^s}$$

which may be hypo-, iso- or hypertonic for a given choice of independent variables. For studying strictly isotonic transport, *TON* is included as independent variable while the diffusion permeability of serosal membrane, P_S^s, is included among the dependent variables:

---

[6] Eq. (13) implies that the compliance factor of the lateral membrane is much smaller than those of apical and serosal membranes. If one instead assumes equal compliance factors of the three plasma membranes it follows that $p^{cell} > p^{bath} = p^o = p^i$. With the input variables of the present computations (Table 3, Fig. 8) and a compliance factor of $1.5 \cdot 10^{-3}$ Pa^{-1} (Spring and Hope 1978), $p^{cell} - p^{bath} = 1.96$ cm H$_2$O with a corresponding increase of C_{ND}^{cell} from 276.96 to 277.04 mM, and all other dependent variables unaffected (see Larsen et al. 2000 for detailed discussion).

$$P_S^s = \frac{TON \cdot (J_V^{bm} + J_V^s) - J_S^{bm}}{C_S^{cell} - C_S^i} \qquad (15)$$

The above equations constitute a closed set and can be solved by numerical methods. With TON set to isotonicity (300 mOsm), P_S^s defines the recirculation flux across the basolateral membrane that would make the net transportate truly isotonic.

The cell volume per unit area of epithelium, V^{cell}, was then calculated from:

$$V^{cell} = \frac{D^{cell} \cdot M_{ND}^{cell}}{C_{ND}^{cell}} \qquad (16)$$

of which the independent variables, D^{cell} and M_{ND}^{cell}, are the number of cells per unit area of epithelium and the amount of non-diffusible solute per cell, respectively. Unidirectional fluxes across tight junction and the interspace basement membrane, respectively, are calculated from:

$$J_S^{tm,\,IN} = J_V^{tm}(1 - \sigma^{tm}) \frac{C_S^o \cdot \exp[J_V^{tm}(1 - \sigma^{tm})/P_S^{tm}]}{\exp[J_V^{tm}(1 - \sigma^{tm})/P_S^{tm}] - 1} + P_S^{tm,\,diff} \cdot C_S^o \quad (17a)$$

$$J_S^{tm,\,OUT} = J_V^{tm}(1 - \sigma^{tm}) \frac{C_S^{lis}}{\exp[J_V^{tm}(1 - \sigma^{tm})/P_S^{tm}] - 1} + P_S^{tm,\,diff} \cdot C_S^{lis} \quad (17b)$$

$$J_S^{bm,\,IN} = J_V^{bm}(1 - \sigma^{bm}) \frac{C_S^{lis} \cdot \exp[J_V^{bm}(1 - \sigma^{bm})/P_S^{bm}]}{\exp[J_V^{bm}(1 - \sigma^{bm})/P_S^{bm}] - 1} \qquad (17c)$$

$$J_S^{bm,\,OUT} = J_V^{bm}(1 - \sigma^{bm}) \frac{C_S^i}{\exp[J_V^{bm}(1 - \sigma^{bm})/P_S^{bm}] - 1} \qquad (17d)$$

With these equations, the paracellular unidirectional fluxes of S are given by:

$$J_S^{para,\,IN} = \frac{(J_S^{tm,\,IN} + J_S^{tm,\,diff,IN}) \cdot J_S^{bm,\,IN}}{J_S^{tm,\,OUT} + J_S^{tm,\,diff,OUT} + J_S^{bm,\,IN}} \qquad (17e)$$

$$J_S^{para,\,OUT} = \frac{J_S^{bm,\,OUT} \cdot (J_S^{tm,\,OUT} + J_S^{tm,\,diff,OUT})}{J_S^{bm,\,IN} + J_S^{tm,\,OUT} + J_S^{tm,\,diff,OUT}} \qquad (17f)$$

4.3
Model Computations: The Unperturbed Intestinal Mucosa

Here we discuss model computations simulating results obtained in our experiments of toad intestine exposed on either side to isotonic Ringer's solution. Specifically, we search for a mathematical solution containing unidirectional electroneutral fluxes of the driving solute, S, that are compa-

Table 3. Values of independent variables used for computations of the model epithelium shown in Fig. 8. Hydraulic permeability of the lateral membrane is here set to zero, so that water is flowing from luminal (o) to lateral intercellular space (lis) through convection-diffusion pores of tight junctions, only

---

INTENSIVE VARIABLES

C_S^o	300 mM	C_S^i	300 mM
C_{ND}^o	0 mM	C_{ND}^i	0 mM

MEMBRANE PARAMETERS, SOLUTES

P_S^a	$3.0 \cdot 10^{-6}$ cm·s^{-1}	P_S^s	$6.7967 \cdot 10^{-6}$ cm·s^{-1}
$J_S^{max, pump}$	$8.3 \cdot 10^3$ pmol·s^{-1}·cm^{-2}	(n = 3 sites on pump)	
P_S^{tm}	$1.5 \cdot 10^{-6}$ cm·s^{-1}	P_S^{bm}	$3.5 \cdot 10^{-5}$
$P_S^{tm. diff}$	0 cm·s^{-1}		

MEMBRANE PARAMETERS, WATER

L^a	$6.0 \cdot 10^{-7}$ (cm·s^{-1})·(N·cm^{-2})$^{-1}$	L^s	$6.0 \cdot 10^{-7}$ (cm·s^{-1})·(N·cm^{-2})$^{-1}$
L^{lm}	0 (cm·s^{-1})·(N·cm^{-2})$^{-1}$		
L^{tm}	$6.0 \cdot 10^{-7}$ (cm·s^{-1})·(N·cm^{-2})$^{-1}$	L^{bm}	$7.5 \cdot 10^{-5}$(cm·s^{-1})·(N·cm^{-2})$^{-1}$
σ^{tm}	0.60	σ^{bm}	10^{-5}

---

rable with the experimental unidirectional fluxes of the charged sodium ion. Solute-pump constants are selected to fulfill the additional requirement that the cellular concentration of S accounts for 5–10% of the osmotic concentration of the cell. Since immunocytochemical studies of the distribution of aquaporins in vertebrate epithelia failed to detect water channels in small intestine (Nielsen et al. 1993), in the computations discussed here we have assumed that water flows from the luminal solution into the lateral intercellular space through convection-diffusion pores of tight junctions. This is being modelled by setting the osmotic water permeability of the *lis* membrane to zero[7]. Finally, in agreement with *in vitro* studies of rat small intestine (Curran 1960) we will assume an isotonic transportate by including the permeability of the serosal membrane, P_S^s, among the dependent variables [*Eq.* (15)] and setting $TON = 300$ mM. This number is equal to the concentration of S of both luminal and serosal bath[8]. A set of independent variables providing the type of mathematical solution searched for is given in Table 3, and the corresponding dependent variables are given in Fig. 8. It is obvious that with a total of 23 independent variables and only 5 dependent variables measured (not counting the cellular [S]), the set of computed results of Fig. 8 is just one out of many that would be reasonably consistent with experimental data. In the theoretical paper (Larsen et al. 2000) a larger number of mathematical solutions were investigated, and it was concluded that the qualitative behaviour of the model is not critically dependent on the choice of independent variables. In other words the computed result discussed here (Fig. 8) is a robust mathematical solution to our problem.

The net flux of the driving solute given by the model is, $\Delta M_S = 1318$ pmol·s^{-1}·cm^{-2}, which should be compared with $\Delta M_{Na} = 773 \pm 56$ pmol · s^{-1}·cm^{-2} of the intestinal preparation. The paracellular fluxes of the model are, $J_S^{para, IN} = 748$ and $J_S^{para, OUT} = 204$ pmol·s^{-1}·cm^{-2}, respectively, with a ratio of, $J_S^{para, IN} / J_S^{para, OUT} = 3.66$. These numbers are to be compared with the intestinal values of, $J_{Na}^{para, IN} = 450 \pm 40$ pmol·s^{-1}·cm^{-2}, $J_{Na}^{para, OUT} = 130 \pm 20$ pmol · s^{-1}·cm^{-2}, and $J_{Na}^{para, IN} / J_{Na}^{para, OUT} = 3.66$, respectively. Thus, the model readily simulates a paracellular Na$^+$ flux-ratio above unity under conditions of transepithelial thermodynamic equilibrium, and ascribes this asymmetry of passive paracellular fluxes to convection fueled by solute pumps on the membranes lining the lateral intercellular space. These fluxes are obtained

---

[7] The original paper (Larsen et al. 2000) also discusses versions of the model with lateral water permeability generating translateral water flows either in combination with junctional water transport, or with water flowing entirely via cells. See also computations presented in Table 5 of the present review.
[8] In these calculations non-diffusible solutes are not added to the external compartments (bathing solutions).

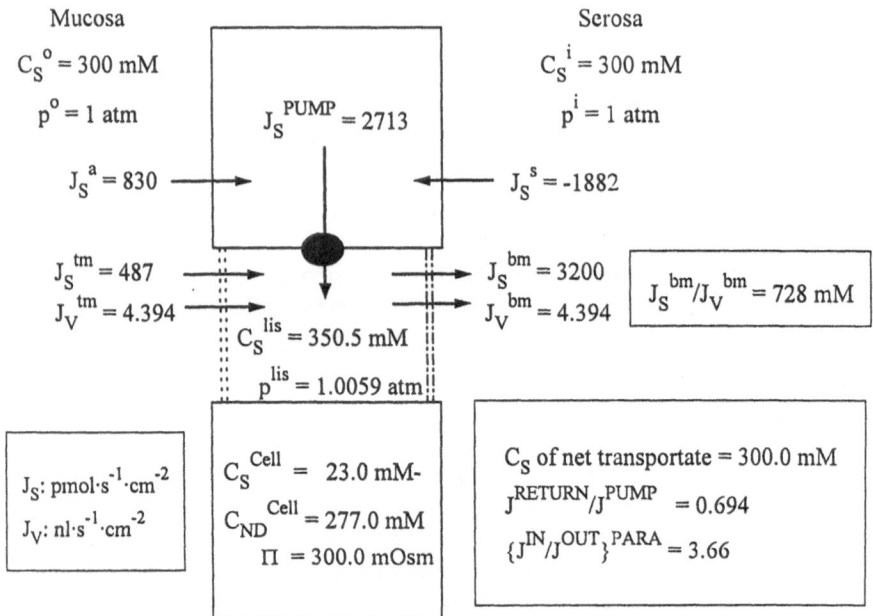

Mucosa

$C_S{}^o = 300$ mM

$p^o = 1$ atm

$J_S{}^{PUMP} = 2713$

Serosa

$C_S{}^i = 300$ mM

$p^i = 1$ atm

$J_S{}^a = 830 \longrightarrow$

$\longleftarrow J_S{}^s = -1882$

$J_S{}^{tm} = 487$

$J_V{}^{tm} = 4.394$

$C_S{}^{lis} = 350.5$ mM

$p^{lis} = 1.0059$ atm

$J_S{}^{bm} = 3200$

$J_V{}^{bm} = 4.394$

$J_S{}^{bm}/J_V{}^{bm} = 728$ mM

J_S: pmol·s^{-1}·cm^{-2}

J_V: nl·s^{-1}·cm^{-2}

$C_S{}^{Cell} = 23.0$ mM-

$C_{ND}{}^{Cell} = 277.0$ mM

$\Pi = 300.0$ mOsm

C_S of net transportate = 300.0 mM

$J^{RETURN}/J^{PUMP} = 0.694$

$\{J^{IN}/J^{OUT}\}^{PARA} = 3.66$

Fig. 8. Computer model of unperturbed toad intestinal mucosa. Note that the virtual concentration of the fluid emerging downstream the lateral space ($J_s{}^{bm}/J_v{}^{bm}$) is significantly larger than that of the lateral intercellular space (conf. analysis of Fig. 7). Isotonicity of the net transportate is maintained by solute recirculation via cells. In this way isotonicity can be maintained at low reflection coefficient of the interspace basement membrane (bm with large pores) and therefore a small excess hydrostatic pressure of the space between cells

with the solute concentration of the lateral space being 20% above that of the bathing solutions. Likewise, the excess hydrostatic pressure of *lis* of 5.9 cm H_2O (Fig. 8) seems to be within acceptable values for biological preparations. It can also be seen that with the paracellular flux-ratio of the model simulating that of the intestinal preparation, the computed virtual concentration of the fluid emerging downstream *lis* is, $J_S{}^{bm}/J_V{}^{bm} = 728$ mM (Fig. 8). This result confirms the conclusion of the basic analysis carried out in Fig. 6, *viz.*, with a relatively small ratio of paracellular fluxes the virtual concentration of the fluid emerging from *lis* is to be significantly larger than the concentration of the fluid in the lateral intercellular space.

Thus, with requirement of isotonic transport our computations predict a fairly large recirculation of *S*, see Fig. 8. In this example with a model pump flux, $J_S{}^{pump} = 2713$ pmol·s^{-1}·cm^{-2}, isotonic transport would require a return flux, $J_S{}^s = -1882$ pmol·s^{-1}·cm^{-2} corresponding to a recirculation of,

$-J_S^s / J_S^{pump} = 0.694$. In experiments with the intestinal preparation we estimated a recirculation flux of, $-J_{Na}^s / J_{Na}^{pump} = 0.65 \pm 0.03$ (listed in Table 2). This agreement between the experimental value of recirculation and the value computed under the assumption of an overall isotonic transport indirectly confirms the hypothesis of Nedergaard et al. (1999), that the physiological significance of ion recirculation is to render the net transportate isotonic. Although this is an encouraging result, it is should be noted that in our studies the intestinal water flux was not directly measured (but indirectly demonstrated by our flux ratio analysis of paracellular markers, conf. *Section 3.1.2*). Therefore, it is not known whether the transportate was truly isotonic, or just close to isotonicity. A more detailed analysis showed that fairly large recirculation fluxes, *i.e.* with $-J_S^s / J_S^{pump}$ in the range of 0.55–0.75, would be required even if the tonicity of the net transportate deviates as much as ±20% from the tonicity of the bathing solutions (Larsen et al. 2000). Thus, the conclusions derived above from computations assuming truly isotonic transport are to be generalized to cases of near isotonic transport.

5
Some Predictions and Possible Applications of the Na⁺-Recirculation Theory

In the concluding sections we discuss more features of the recirculation model for investigating the explanatory range of the theory. Rather than going through detailed computations simulating the large number of leaky epithelia that have been investigated experimentally, we will confine the analysis to a few examples. They have been selected for pointing out that leaky epithelia not only share capacity for isotonic transport, but that a number of other features characterize epithelia specialized for solute coupled water transport, such as proportionality between solute transport and volume flow, apparently unpredictable energy expenditure of active sodium transport, and transport of water uphill. A satisfactory theoretical description would have to deal in a logical way with all of these features.

5.1
Relationship Between Solute Transport and Volume Flow

In Curran's classical study it was observed that the volume flow across the intestinal preparation is increasing in a linear fashion with the net flux of Na⁺, and that the volume flow is zero when the net flux of Na⁺ is zero

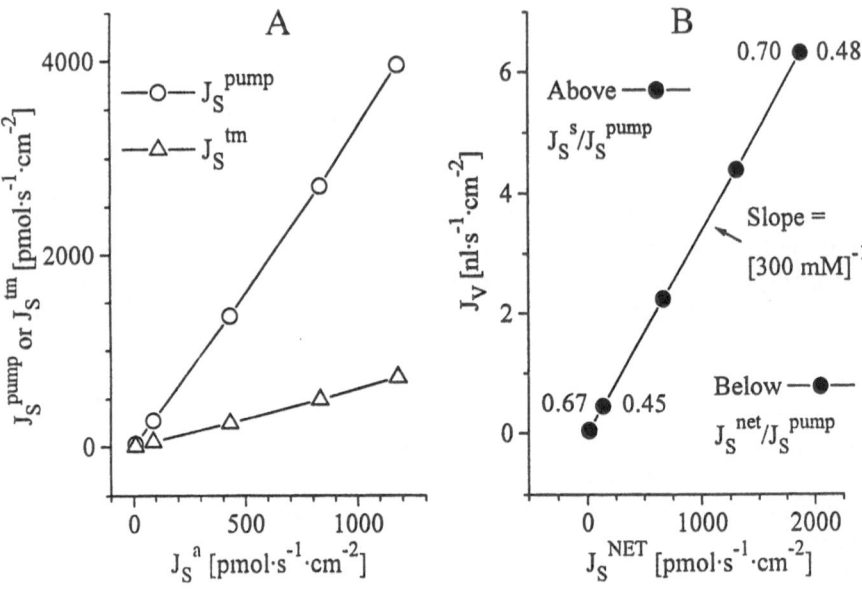

Fig. 9. A. Relationship between cellular uptake of S from the luminal solution and the lateral pump flux and the convection-diffusion flux through tight junctions, respectively. J_S^a, was varied by varying the apical solute permeability, P_S^a, resulting in a variation of the cellular concentration of S from 1.77 to 37.9 mM. **B.** Since water is driven by the solute pumps, in the recirculation model is volume flow proportional with the transmural net transport of the driving solute, similar to what Curran (1960) found for rat small intestine. Isotonicity is maintained at the expense of energetic efficiency of solute transport. Thus the net transepithelial flux of the solute amounts to no more than half of the pumped flux (numbers of J_S^{net}/J_S^{pump} are indicated below the graph while the relative recirculation flux, J_S^s/J_S^{pump}, is indicated above the graph). A small energetic efficiency was predicted from fluxes measured in toad intestine (Table 2). In the text are given examples of energetic efficiencies exceeding that of the pump itself, and it is discussed that also this regime is explained by the recirculation model

(Curran 1960, *loc. cit.* Fig. 1). Such a relationship is given by the model, see Fig. 9B. In these computations, was the net transport of S varied by varying the solute permeability of the luminal plasma membrane, P_S^a. The cellular uptake of S, J_S^a, increases with increasing apical diffusion permeability. In Fig. 9A is shown that the pump flux, J_S^{pump}, and the convection-diffusion flux of S through tight junction, J_S^{tm}, respectively, increase linearly with J_S^a (C_S^{cell} is far from saturating the lateral pumps). The latter result follows from the fact that the water flux through tight junctions depends on the activity of the lateral solute pumps. The slope of the J_S^{pump}/J_S^a-relationship is much above one illustrating the significant recirculation flux. The resulting linear dependence of volume flow on net solute transport is shown in Fig. 9B. The conclusion is that in the recirculation model the flux of S is rate limiting for the transepithelial flow of water because it is the lateral solute pumps that drive water through the epithelium. At all transport rates the relative recirculation flux is about the same (*circa* 0.7) with an energetic efficiency of the solute transport being low ($J_S^{net} / J_S^{pump} \ll 1$). We will discuss this result in the section below.

5.2
Active Transport and Energy Expenditure

The experimental results given in Tables 1 and 2 above (*Section 3.3.2*) implies that in toad small intestine is the Na^+-pump flux significantly larger than the net transepithelial flux of this ion. The pumped fluxes directed to the mucosal and serosal bath, respectively, are calculated from (see Fig. 5):

$$\sum J_{ms}^{bm} = J_{ms}^{active} + J_{sm}^{active, return} \tag{18a}$$

$$\sum J_{sm}^{tm} = J_{sm}^{active} + J_{ms}^{active, return} \tag{18b}$$

The component given by *Eq.* (18b) represents a leak of the system. By being dependent on the cation permeability of tight junction, it may well vary considerably among different types of epithelia[9]. The pumped fluxes given by *Eqs.* (18) are collected in Table 4, which also contains the experimentally measured net inward fluxes. In most studies the cost of isotonic transport has been determined by measuring net inward transport of Na^+ together with the O_2-consumption associated with the working of the sodium pump. The pump itself would transport ~18 mol Na^+/mol O_2, and this is the number usually arrived at in studies on tight epithelia (Zerahn 1956, Leaf and Renshaw 1957, Leaf and Dempsey 1960). For toad small intestine the expected oxygen consumption would be 4.5 Na^+/O_2 (Table 4, 6th column). The relatively small number (4.5 as compared to 18 for the pump) is of course due to the significant recirculation flux. A similar small energetic efficiency of net Na^+ transport was observed in the frog skin following reversible 'opening up' of the tight junctions by exposure of the apical plasma membranes to a hypertonic solution (Ussing 1966). Under some experimental conditions gallbladder epithelium also exhibits a relatively low energetic efficiency (13 Na^+/O_2, Frederiksen and Leyssac 1969), whereas under other

---

[9] The presence of such a leak is a prerequisite for Na^+ flowing along the paracellular pathway, and only in epithelia with a finite Na^+ permeability of tight junction is the analysis presented in the experimental paper (Nedergaard et al. 1999) possible.

Table 4. Toad small intestine. Estimation of the cost of isotonic transport by toad small intestine. Experimental fluxes are taken from Table 1[a]. Calculations were carried out for each experiment with their mean values and standard errors indicated (fluxes in $pmol \cdot s^{-1} \cdot cm^{-2}$). Right hand column: Estimated oxygen consumption per sodium ion transported across the preparation (mole Na^+/mole O_2) assuming an ATP:O_2-ratio = 6

$\Sigma J_{ms}^{active,bm}$ (1)	$\Sigma J_{sm}^{active,tm}$ (2)	Pump Flux (1) + (2)	$\Delta M_{Na} =$ $M_{ms} - M_{sm}$	ΔM_{Na}/ATP	$\Delta M_{Na}/O_2$
2616 ±401	683 ±145	3290 ±486	775 ±55	0.75 ±0.09	4.5

[a] $\Sigma J_{ms}^{active,bm} = J_{ms}^{active} + J_{sm}^{active,return}$ $\Sigma J_{ms}^{active,tm} = J_{sm}^{active} + J_{ms}^{active,return}$

conditions the gallbladder exhibits efficiencies as large as 25–30 Na^+/O_2 (Martin and Diamond 1966, Frederiksen and Leyssac 1969). Similar large numbers, about 25 Na^+/O_2, have been reported in studies of kidney proximal tubule (Deetjen and Kramer 1961, Lassen and Thaysen 1961, Lassen et al. 1961, Thurau 1961, Torelli et al. 1966). Taken together this set of different numbers of energy expenditure obtained in early studies of leaky epithelia represents a true paradox, which has never been resolved. According to the concept presented above, the low energetic efficiency would be consistent with a relatively large ("invisible") recirculation flux, whereas efficiency over and above that of the pump itself might be associated with a relative large paracellular convection flux combined with a small recirculation flux. In what follows we investigate whether the range of costs discussed above are to be predicted from more rigorous computer assisted analyses of the recirculation model.

With three examples, the results of these computations are listed in Table 5. In the 2nd column are shown numbers of relevance for the discussion of the unperturbed mucosa given in Fig. 8. According to these computations the energetic efficiency is no more than half that of the pump itself[10]. In the 3rd column is shown a similar set of calculations but with the hydraulic

---

[10] This is more than estimated directly from measured fluxes (Table 4), and is due to the fact that the model in its main outline, but not in details, simulates the experimental preparation. Perhaps not unexpected since the model deals with electroneutral solute transport, and not transport of ions.

Table 5. Model calculations of energy expenditure of solute coupled water transport in a leaky epithelium with recirculation of the driving species. The three examples listed corresponds to, *2nd column*: the intestinal mucosa with transjunctional water flow (conf. Table 3 and Fig. 8), *3rd column* the transjunctional water permeability is increased by multiplying L^{tm} and L^{bm} by a factor of ten, and in the *4th column* the water permeability of the translateral route is increased to a similar value allowing water to move through cells before entering the lateral space*

Dependent variables	$\dfrac{L^{tm}\cdot L^{bm}}{L^{tm}+L^{bm}}$ $= 5.95\cdot 10^{-7}$ $(cm\cdot s^{-1})/(N\cdot cm^{-2})$	$\dfrac{L^{tm}\cdot L^{bm}}{L^{tm}+L^{bm}}$ $= 5.95\cdot 10^{-6}$ $(cm\cdot s^{-1})/(N\cdot cm^{-2})$	$\dfrac{L^{cell}\cdot L^{bm}}{L^{cell}+L^{bm}}$ $= 5.95\cdot 10^{-6}$ $(cm\cdot s^{-1})/(N\cdot cm^{-2})$
J_S^{pump} (pmol·s^{-1}·cm^{-2})	2713	1075	1018
J_S^{net} (pmol·s^{-1}·cm^{-2})	1318	1441	903.2
J_V^{net} (nl·s^{-1}·cm^{-2})	4.394	4.802	3.011
C_S^{cell} (mmol/l)	23.0	10.5	10.1
$C_S^{cell} + C_{ND}^{cell}$ (mmol/l)	300	300	301.8
C_S^{lis} (mmol/l)	350.5	305.5	303.7
J_S^{tm} (pmol·s^{-1}·cm^{-2})	487	572	33.5
$J_S^{para, IN} / J_S^{para, OUT}$	3.66	4.13	1.19
J_S^{bm} / J_V^{bm} (mmol/l)	728	343	345
J_S^{s} / J_S^{pump}	0.694	0.192	0.146
J_S^{net} / J_S^{pump}	0.486	1.340	0.887

* $L^{cell} = \dfrac{L^a \cdot L^{lis}}{L^a + L^{lis}}$ with $L^a = L^{lis} = 6\cdot 10^{-6}$, $L^{bm} = 7.5\cdot 10^{-4}$ (cm·s^{-1})/(N·cm^{-2}) and L^s set to such a small value that water recirculation across the inner border of the epithelium is to be disregarded. All other independent variables kept at their reference values.

conductance of the paracellular pathway increased by a factor of 10. The 4th column contains a data set obtained by increasing the hydraulic conductance of the translateral pathway to a similar high value (see legend of Table 5). In both cases, the recirculation flux is decreased and the overall energetic efficiency is increased. The following reasoning explains this result. With a much larger hydraulic conductance the driving force ($\approx C_S^{lis} - C_S^o$) needed to generate the stationary flow of water from the luminal solution to the coupling compartment is smaller. At the new steady state, therefore, the pump flux is smaller and, *pari passu*, so is the steady state C_S^{cell}. This leads to the expected [S]-decrease of *lis* with a correspondingly smaller virtual concentration of the transportate emerging from *lis*. Accordingly, the recirculation flux necessary to render the net transportate isotonic is reduced.

The interesting point is that the energetic efficiency of solute coupled water transport is governed by the water permeability of the pathway carrying water from the external (luminal) to the coupling (lateral) compartment no matter whether water takes a transjunctional or a translateral route. By comparing the results obtained with predominant transjunctional and translateral water flow, respectively, the significance of tight junction convection flow for the energetic efficiency is illustrated. In the case of a relatively large water flow through convection-diffusion pores of tight junctions, the direct uptake of solute from the luminal compartment is large (572 pmol·s^{-1}·cm^{-2}) resulting in an energetic efficiency over and above that of the pump itself ($J_S^{net} / J_S^{pump} = 1.340$; Table 5 column 3, corresponding to 24 Na$^+$/O$_2$). With water flowing predominantly through cells, the paracellular convection flux is much smaller (33.5 pmol·s^{-1}·cm^{-2}) and so is the efficiency of net solute transport ($J_S^{net} / J_S^{pump} = 0.887$; Table 5 column 4).

We are now prepared for proposing a configuration of solute and hydraulic permeabilities, which would lead to the very large volume flows and energetic efficiencies of mammalian proximal tubule. Whereas amphibian epithelia exhibits solute coupled volume flows in order of 1–10 nl·s^{-1}·cm^{-2}, rat proximal tubule generates isotonic water flows of more than 50 nl·s^{-1}·cm^{-2} (Weinstein 1992). Since the rate of volume flow is governed by the rate of active transport, first we would need to increase the rate of solute pumping. For maintaining reasonably cellular [S] we would have to increase P_S^a as well as $J_S^{pump,max}$. This maneuver by itself would lead to an increased rate of solute coupled water transport. However, for obtaining high efficiency of transport the water permeability would have to be increased as well. The computations are discussed in the original model paper (Larsen et al. 2000) showing that with a low lateral intercellular solute concentration (< 306 mM) both the high rate of volume flow and the high energetic efficiency of net solute transport are being simulated by the model.

5.3
Uphill Water Transport and Anomalous Solvent Drag

Leaky epithelia have capacity of transporting water against an adverse osmotic gradient. This was demonstrated by Grim and Smith (1957) and by
Parsons and Wingate (1958) in their studies on dog gallbladder and rat jejunum, respectively. The conclusion of these early studies has been generalized for fish (Diamond 1962a), rabbit (Diamond 1964a), and *Necturus*
(Persson and Spring 1982) gallbladder, small intestine of dog (Hakim et al.
1963) and rat (Pappenheimer and Reis 1987), and more recently also for
proximal tubule of rat kidney (Green et al. 1991). In our studies we observed
paracellular convection of radiolabelled cesium ions against an adverse
osmotic concentration difference of at least 150 mM (Fig. 3). This finding
indicates that the mechanism of uphill water transport depends on paracellular coupling between the water flux and active Na^+ flux, *i.e.*, similar to the
mechanism of isotonic water transport. If this suggestion holds true there is
no need to postulate the operation of a cellular water pump, as metabolic
energy for the above mechanism couples to the Na^+/K^+-ATPase of the membranes lining the lateral intercellular space. This hypothesis was investigated
by analysis of a compartment model of the lateral intercellular space of
proximal tubule (Weinstein and Stephenson 1981b). It was concluded that
by raising the solute concentration of the lateral intercellular space above
that of the luminal solution, the solute pump would have capacity for generating a downhill flux of water into this space. Fluid would then be forced
into the serosal compartment by the hydrostatic pressure difference build
up between these two compartments. For the recirculation model we also
verified uphill water transport (Larsen et al. 2000), but unlike the Weinstein-
Stephenson model, the recirculation model also produces a truly isotonic
transportate under thermodynamic equilibrium conditions (Larsen et al.
2000, see also Fig. 8 and Table 5 of the present review).

The concept of anomalous solvent drag was introduced in a study of the
leaky frog skin in which it was shown that paracellular marker molecules
were dragged in the inward direction in spite of a prevailing flux of water in
the outward direction (Ussing 1966, Franz and Van Bruggen 1967). With a
hydraulic conductance of the apical and serosal membrane water can flow
both via the cellular and the paracellular pathway. With this configuration
the recirculation model exhibit anomalous solvent drag (Larsen et al. 2000).
The mechanism is as follows. With a raised osmotic pressure of the luminal
solution water flows in the outward direction via the cells driven by the established transepithelial osmotic gradient, and in the inward direction via
the paracellular pathway energized by the lateral solute pumps. In the range

of osmotic driving forces where the backward flux of water via cells is larger than the forward flux of water via the coupling compartment, paracellular markers are driven by convection in the inward direction, while the net flow of water is outward.

Our computations showed that isotonic transport, uphill water transport, and anomalous solvent drag could be obtained with a common set of independent variables. Thus with no additional assumptions the recirculation theory handles all three types of water transport characterizing leaky epithelia.

Acknowledgemens. Scientific research in the authors' laboratories is supported by grants from the Danish Natural Science Research Council (11–0971), and from the Alfred Benzon, Carlsberg, and Novo Nordisk Foundations. In preparing the manuscript, technical assistance by Annette Berthelsen, Anni Olsen, Birthe Petersen, and Thomas Sørensen is greatly appreciated.

References

Agre P (1998) Aquaporin null hypothesis:The importance of classical physiology. Proc Natl Acad Sci USA 95:9061–9063

Agre P, Nielsen S (1996) The aquaporin family of water channels in kidney. Nephrology 17:409–415

Andreoli TE, Schafer JA (1978a) Volume absorption in the pars recta. III. Luminal hypotonicity as a driving force for isotonic volume absorption. Am J Physiol 234:F349–F355

Andreoli TE, Schafer JA (1978b) External solution driving forces for isotonic volume absorption. Fed Proc 38:154–160

Andreoli TE, Schafer JA (1979) Effective luminal hypotonicity: the driving force for isotonic proximal tubular fluid absorption Am J Physiol 236:F89–F96

Andreoli TE, and Schafer JA, Troutman SL, Watkins ML (1979) Solvent drag component of Cl⁻ flux in superficial proximal straight tubules:evidence for a paracellular component of isotonic fluid absorption. Am J Physiol 237:F455–F462

Atisook K, Carlson S, Madara JL (1990) Effects of phlorizin and sodium on glucose-elicited alterations of cell junctions in intestinal epithelia. Am J Physiol 258:C77–C85

Barry HB, Diamond JM (1984) Effects of unstirred layers on membrane phenomena Physiol Rev 64:763–872

Berridge MJ, Gupta BL (1967) Fine-structural changes in relation to ion and water transport in the rectal papillae of the blowfly. *Calliphora.* J Cell Sc 2:89–112

Boucher RC (1994a) Human airway ion transport. Part One. Am J Respir Crit Care Med 150:271–281

Boucher RC (1994a) Human airway ion transport. Part Two. Am J Respir Crit Care Med 150:581–593

Boucher RC (1999) Molecular insights into the physiology of the 'thin film' of airway surface liquid. J Physiol (London) 516:631–638

Boucher RC, Larsen EH (1988) Comparison of ion transport by cultures of secretory and absorptive canine airway epithelia. Am J Physiol 254:C535–C547

Boulpaep EL (1971) Electrophysiological properties of the proximal tubule: Importance of cellular and intercellular transport pathways. In: Giebisch G (ed) Electrophysiology of epithelial cells. Scattauer, Stuttgart New York, pp 91–118

Boulpaep EL, Maunsbach AB, Tripathi S, Weber, MR (1993) Mechanism of isosmotic water ransport in leaky epithelia:Consensus and inconsistencies. In: Ussing HH, Fischbarg J, Sten-Knudsen O, Larsen EH, Willumsen NJ (eds) Isotonic Transport in Leaky Epithelia Proc. Alfred Benzon Symp. 34 Munksgaard, Copenhagen, pp 53–67

Carpi-Medina P, Whittembury G (1988) Comparison of transcellular and transepithelial water osmotic permeability (P_{os}) in isolated proximal straight tubule (PST) of the rabbit kidney. Pflügers Arch 412:66–74

Chatton J-Y, Spring KR (1995) The sodium concentration of the lateral intercellular space of MDCK cells: A microspectrofluorimetric study. J Membrane Biol 144:11–19

Cotton CU, Weinstein, AM, Reuss L (1989) Osmotic water permeability of *Necturus* gallbladder. J Gen Physiol 93:649–679

Curran PF (1960) Na, Cl, and water transport by rat ileum *in vitro*. J Gen Physiol 43:1137–1148

Curran PF, Macintosh J.R (1962). A model system for biological water transport. Nature 193:347–348

Curran PF, Solomon AK (1957) Ion and water fluxes in the ileum of rats. J Gen Physiol 41:143–168

Dawson DC, Smith SS, Mansoura MK (1999) CFTR:Mechanism of anion conduction. Physiol Rev 79:S47-S75

Deetjen P, Kramer K (1961) Die Abhängigkeit des O_2-Verbrauchs der Niere von der Na-Rückresorption. Pflügers Arch 273:636–642

Diamond JM (1962a) The reabsorptive function of the gall-bladder. J Physiol 161:442–473

Diamond JM (1962b) The mechanism of solute transport by the gall-bladder. J Physiol 161:474–502

Diamond JM (1964a) Transport of salt and water in rabbit and guinea pig gall bladder. J Gen Physiol 48:1–14

Diamond JM (1964b) The mechanism of isotonic water transport. J Gen Physiol 48:15–42

Diamond JM, Bossert WH (1967) Standing gradient osmotic flow. A mechanism for coupling of water and solute transport in epithelia. J Gen Physiol 50:2061–2083

Diamond JM, Bossert WH (1968). Functional consequences of ultrastructural geometry in "backwards fluid-transporting epithelia". J Gen Physiol. 37:694–702

DiBona DR, Mills JW (1979) Distribution of Na-pump sites in transporting epithelia. Fed Proc 38:134–143

Diener M, Rummel W, Mestres P, Lindemann B (1989) Single chloride channels in colon mucosa and isolated colonic enterocytes of the rat. J Membrane Biol 108:21–30

Ecke D, Bleich M, Greger R (1996) Crypt base cells show forskolin-induced Cl⁻ secretion but no cation inward current. Pflügers Arch 431:427–434

Field M (1971) Ion transport in rabbit ileal mucosa. II. Effects of cyclic 3', 5'-AMP. Am J Physiol, 221:992–997

Field M (1974) Intestinal secretion. Gastroenterology 66:1063–1084

Field M., Plotkin GR, Silen W (1968). Effects of vasopressin, theophylline and cyclic adenosine monophosphate on short circuit current across isolated ileal mucosa. Nature 217:469–471

Fischbarg J, Kuang K, Hirsch J, Lecuona S, Rogozinski L, Silverstein SC, Loike J (1989). Evidence that the glucose transporter serves as a water channel in J774 macrophages. Proc Natl Acad Sci USA 86:8397–8401

Fischbarg J, Kuang K, Li J, Vera JC (1990). Glucose transporters serve as water channels. Proc Natl Acad Sci USA 87:3244–3247

Fischbarg J, K. Kuang, J. Li, Arant-Hickman S, Vera JC, Silverstein SC, Loike JD (1993) Facilitative and sodium-dependent glucose transporters behave as water channels. In: Isotonic Transport in Leaky Epithelia. Eds.: Ussing HH, Fischbarg J, Sten-Knudsen O, Larsen EH, Willumsen NJ. Proc. Alfred Benzon Symp 34:432–446, Munksgaard, Copenhagen

Fordtran JS, Dietschy JM (1966) Water and electrolyte movement in the intestine. Gastroenterology 50:263–285

Fordtran JS, Rector FC, Carter NW (1968) The mechanism of sodium absorption in the human small intestine. J Clin Invest 47:884–900

Forsyth GW, Gabriel SE (1989) Activation of chloride conductance in pig jejunum brush border vesicle. J Membrane Biol 107:137–144

Franz TJ, Van Bruggen JT (1967) Hyperosmolarity and the net transport on non-electrolyes in frog skin J Gen Physiol 50:933–949

Frederiksen O, Leyssac PP (1969) Transcellular transport of isosmotic volumes by the rabbit gall-bladder in vitro. J Physiol 210:201–224

Frederiksen O, Leyssac PP (1974) Absence of dilated lateral intercellular spaces in fluid-transporting frog gallbladder epithelium. Direct microscopy observations. J Cell Biol 61:830–834

Frizzell RA, Morris AP (1994) Chloride conductance of salt secreting epithelial cells. Curr Top Membr 42:173–214

Frizzell RA, Nellans HN, Rose RC, Markscheid-Kaspi L, Schultz SG (1973) Intracellular Cl concentration and influxes across the brush border of rabbit ileum. Am J Physiol 224:328–337

Frizzell RA, Schultz SG (1972) Ionic conductances of extracellular shunt pathway in rabbit ileum. Influence of shunt on transmural sodium transport and electrical potential. J Gen Physiol. 59:318–346

Frizzell RA, Field M, Schultz SG (1979) Sodium-coupled chloride transport by epithelial tissues. Am J Physiol 236:F1–F8

Frizzell RA, Smith PL, Vosburgh E, Field M (1979) Coupled sodium-chloride influx across brush border of flounder intestine. J Membrane Biol 46:27–39

Frömter E (1972) The route of passive ion movement through the epithelium of Necturus gallbladder. J Membrane Biol 8:259–301

Frömter E (1973) The role of terminal bars for transepithelial current flow through Necturus gallbladder epithelium. In:Ussing HH, Thorn NA (eds) Transport Mechanisms in Epithelia. Proc. Alfred Benzon Symp. V:492–504, Munksgaard, Copenhagen

Frömter E (1979) Solute transport across epithelia: What can we learn from micropuncture studies of kidney tubules. J Physiol (London) 288:1–31

Frömter E (1986) The electrophysiological analysis of tubular transport. Kidney Int. 30:216–228

Frömter E, Diamond JM (1972) Route of passive ion permeation in epithelia. Nature New Biology, 235:9–13

Frömter E, Müller CW, Wick T (1971) Permeability properties of the proximal tubular epithelium of the rat kidney studied with electrophysiological methods. In: Giebisch G (ed) Electrophysiology of Epithelial Cells. F.K. Schattauer Verlag, Stuttgart, pp 119–148

Frömter E, Rumrich G, Ullrich KJ (1973) Phenomenologic description of Na^+, Cl^- and HCO_3^- absorption. Pflügers Arch 343:189–220

Gadsby DC, Nairn AC (1999) Control of CFTR gating by phosphorylation and nucleotide hydrolysis. Physiol Rev 79:S77–S107

Gamba G, Saltzberg SN, Lombardi M, Miyanoshita A, Lytton J, Hediger MA,Brenner BM, Hebert SC (1993) Primary structure and functional expression of a cDNA encoding the thiazide-sensitive electroneutral sodium-chloride cotransporter. Proc Natl Acad Sci USA 90:2749–2753

Giraldez F, Sepulveda FV (1987) Changes in the apparent chloride permeability of *Necturus* enterocytes during sodium-coupled transport of alanine. Biochim Biophys Acta 898:248–252

Giraldez F, Murray KJ, Sepulveda FV, Sheppard DN (1989) Characterization of a phosphorylation-activated Cl^- selective channel in isolated *Necturus* enterocytes. J Physiol 416:517–537

Giraldez F, Sepulveda FV, Sheppard DN (1988) A chloride conductance activated by adenosine 3',5'-cyclic monophosphate in the apical membrane of *Necturus* enterocytes. J Physiol 395:597–623

Gögelein H (1988) Chloride channels in epithelia. Biochim Biophys Acta 947:521–547

Green R, Giebisch G, Unwin R, Weinstein, A.M (1991) Coupled water transport by rat proximal tubule. Am J Physiol 261:F1046–F1054

Greger R, Bleich M, Leipziger J, Ecke D, Mall M, Kunzelmann K (1997) Regulation of ion transport in colonic crypts. News Physiol Sci 12:62–66

Greger R, Schlatter E (1984a) Mechanism of NaCl secretion in the rectal gland of spiny dogfish (*Squalus acanthias*). I. Experiments on isolated in vitro perfused gland tubules. Pflügers Arch 402:63–75

Greger R, Schlatter E (1984b) Mechanism of NaCl secretion in the rectal gland of spiny dogfish (*Squalus acanthias*). II. Effects of inhibitors. Pflügers Arch 402:364–375

Grim E, Smith GA (1957) Water flux rates across dog gallbladder wall. Am J Physiol 191:555–560

Grubb BR, Boucher RB (1999) Pathophysiology of gene-targeted mouse models for cystic fibrosis. Physiol Rev. 79:S193–S214

Gunter-Smith PJ, Grasset E, Schultz SG (1982) Sodium-coupled amino acid and sugar transport by *Necturus* small intestine. An equivalent electrical circuit analysis of a rheogenic co-transport system. J Membrane Biol 66:25–39

Gupta BL, Hall TA (1981) Microprobe analysis of fluid transporting epithelia: Evidence for local osmosis and solute recycling. In:Water Transport in Epithelia, ed.: Ussing HH, Bindslev N, Lassen NA, Sten-Knudsen O. Proc Alfred Benzon 15, Munksgaard, Copenhagen, 17–35

Hakim A, Leister RG, Lifson N (1963) Absorption by an in vitro preparation of dog intestinal mucosa. J Applied Physiol 18:409–413

Halm, DR, Krasny EJ, Frizzell RA (1985a) Electrophysiology of flounder intestinal mucosa. I. Conductance properties of the cellular and paracellular pathways. J Gen Physiol 85:843–864

Halm, DR, Krasny EJ, Frizzell RA (1985b) Electrophysiology of flounder intestinal mucosa. II. Relation of the electric potential to coupled Na-Cl absorption. J Gen Physiol 85:865–883

Hannafin J, Kinne-Saffran E, Frieman D, Kinne R (1983) Presence of a sodium-potassium chloride cotransport system in the rectal gland of *Squalus acanthias*. J Membrane Biol 75:73–83

Hediger MA, Coady MJ, Ikeda TS, Wright EM (1987) Expression cloning and cDNA sequencing of the Na^+/glucose co-transporter. Nature 330:379–381

Hierholzer K, Kawamura S, Seldin DW, Kokko JP, Jacobson HR (1980). Reflection coefficients of various substances across superficial and juxtamedullary proximal convoluted segments of rabbit nephron. Miner Electrolyte Metab 3:172–180

Hill AE, Hill BS (1978) Sucrose fluxes and junctional water flow across *Necturus* gall bladder epithelium. Proc R Soc Lond B:163–174

Hill AE, Shachar-Hill B (1993) A mechanism for isotonic fluid flow through the tight junctions of *Necturus* gallbladder epithelium. J Membrane Biol 136:253–262

Holtug K, Shipley A, Dantzler V, Sten–Knudsen O, Skadhauge E (1991) Localization of sodium absorption and chloride secretion in an intestinal epithelium. J Membrane Biol 122:215–229

Hopfer U, Liedtke CM (1987) Proton and bicarbonate transport mechanisms in the intestine. Annu Rev Physiol 49:51–67

Ikonomov O, Simon M, Frömter E (1985) Electrophysiological studies on lateral intercellular spaces in *Necturus* gallbladder epithelium. Pflügers Arch 403:301–307

Jacobson HR, Kokko JP, Seldin DW, Holmberg C (1982) Lack of solvent drag of NaCl and $NaHCO_3$ in rabbit proximal tubule. Am J Physiol 243:F342–F348

Kaye GI, Wheeler HO, Whitlock RT, Lane N (1966) Fluid transport in the rabbit gall bladder. A combined physiological and electron microscopic study. J Cell Biol 30:237–268

Knickelbein R, Aronson PS, Schron CM, Seifter J, Dobbins JW (1985) Sodium and chloride transport in rabbit ileal brush border. II. Evidence for $Cl–HCO_3$ exchange and mechanism of coupling. Am J Physiol 249:G236–G245

Kovbasnjuk O, Leader JP, Weinstein AM, Spring KR (1998) Water does not flow across the tight junctions of MDCK cell epithelium. Proc Natl Acad Sci USA 95:6526–6530

Larsen EH, Sørensen JN (1999) A mathematical model of fluid transporting leaky epithelia incorporating recirculation of the driving solute. FASEB J 13:A74, 111.7

Larsen EH, Sørensen JB, Sørensen JN (2000) A mathematical model of solute coupled water transport in toad intestine incorporating active recirculation of the driving species. J Gen Physiol (submitted)

Larson M, Spring KR (1983) Bumetanide inhibition of NaCl transport by *Necturus* gallbladder. J Membrane Biol 74:123–129

Lassen UV, Thaysen JH (1961) Correlation between sodium transport and oxygen consumption in isolated renal tissue. Biophys Biochem Acta 47:616–618

Lassen NA, Munk O, Thaysen JH (1961) Oxygen consumption and sodium reabsorption in the kidney. Acta Physiol Scand 51:371–384

Leaf A, Dempsey, E (1960) Some effects of mammalian neurohypophyseal hormones on metabolism and active transport of sodium by the isolated toad bladder. J Biol Chem 235:2160

Leaf A, Renshaw A (1957) Ion transport and respiration of isolated frog skin. Biochem J 65:82

Liedtke CM, Hopfer U (1982a) Mechanism of Cl⁻ translocation across small intestinal brush-border membrane. I. Absence of Na⁺-Cl⁻ cotransport. Am J Physiol 242:G263–G271

Liedtke CM, Hopfer U (1982b) Mechanism of Cl⁻ translocation across small intestinal brush-border membrane. II. Demonstration of Cl⁻OH⁻ exchange and Cl⁻ conductance. Am J Physiol 242:G272–G280

Loo DDF, Zeuthen T, Chandy G, and Wright EM (1996) Cotransport of water by the Na⁺/glucose cotransporter. Proc Natl Acad Sci USA 93:13367–13370

Loretz CA, Fourtner CR (1988) Functional characterization of a voltage-gated anion channel from teleost fish intestinal epithelium. J Exp Biol 136:383–403

Madara JL (1991) Functional morphology of epithelium of the small intestine. In: Handbook of Physiology Sect 6, Vol. IV, pp 83–120. American Physiological Society, Washington

Martin, DW, Diamond JM (1966) Energetics of coupled active transport of sodium and chloride. J Gen Physiol 50:295–315

Matthews JB, Hassan I, Meng S, Archer A, Hrnjez B, Hodin R (1998) Na-K-2Cl cotransporter gene expression and function during enterocyte differentiation. J Clin Invest 101:2072–2079

Maunsbach AB, Boulpaep EL (1984) Quantitative ultrastructure and functional correlates in proximal tubule of Ambystoma and Necturus. Am J Physiol 246:F710–F724

Maunsbach AB, Boulpaep EL (1991) Immunoelectron microsope localization of Na,K- ATPase in transport pathways in proximal tubule epithelium. Micron Microscop Acta 22:55–56

Mills JW, Dibona DR (1978) Distribution of Na⁺ pump sites in the frog gallbladder. Nature 271:273–275

Musch MW, Orellana SA, Kimberg LS, Field M, Halm D, Krasny EJ, Frizzell RA (1982) Na⁺-K⁺-Cl⁻ co-transport in the intestine of a marine teleost. Nature 300:351–353

Nedergaard S (1999) Sodium recirculation and isotonic water transport across isolated toad intestine. Physiol Res 48 Supl 1:S100

Nedergaard S, Larsen EH, Ussing H.H (1999). Sodium recirculation and isotonic transport in toad small intestine. J Membr Biol 168:241–251

Nellans HN, Frizzell RA, Schultz SG (1973) Coupled sodium-chloride influx across the brush border of rabbit ileum. Am J Physiol 225:467–475

Nielsen R (1997) Correlation between transepithelial Na transport and transepithelial water movement across frog skin (Rana esculenta). J Membrane Biol 159:61–69

Nielsen S, Agre P (1995) The aquaporin family of water channels in kidney. Kidney International 48:1057–1068

Nielsen S, Smith BL, Christensen EI, Agre P (1993). Distribution of aquaporin CHIP in secretory and resorptive epithelia and capillary epithelia. Proc Natl Acad Sci USA 90:7275–7279

O'Grady SM, Musch MW, Field AM (1986) Stoichiometry and ion affinities of the Na-K-Cl cotransport system in the intestine of the winter flounder (*Pseudopleuronectes americanus*). J Membrane Biol 91:33–41

O'Grady SM, Palfrey HC, Field AM (1987) Characteristics and functions of the Na/K/2Cl cotransport in epithelial tissues. Cell Physiol Biochem 2:293–307

Orlowski J, Kandasamy RA, Shull GE (1992) Molecular cloning of putative members of the NHE exchanger gene family. cDNA cloning, deduced amino acid sequence, and messenger RNA tissue expression of the rat NHE exchanger NHE-1 and two structurally related proteins. J Biol Chem 267:9331–9339

Okada Y, Tsuchiya W, Irimajiri A, Inouye A (1977) Electrical properties and active solute transport in rat small intestine. I. Potential profile changes associated with sugar and amino acid transport. J Membrane Biol 31:205–219

Pappenheimer JR (1987) Physiological regulation of transepithelial impedance in the intestinal mucosa of rats. J Membrane Biol 100:137–148

Pappenheimer JR (1993) On the coupling of membrane digestion with intestinal absorption of sugars and amino acids. Am J Physiol 265:G409–G417

Pappenheimer JR, Reiss KZ (1987) Contribution of solvent drag through intercellular junctions to absorption of nutrients by small intestine of the rat. J Membrane Biol 100:123–136

Parent L, Supplison S, Loo DDE, Wright EM (1992a) Electrogenic properties of the cloned Na$^+$/glucose cotransporter: I. Voltage clamp studies. J Membrane Biol 125:49–62

Parent L, Supplison S, Loo DDE, Wright EM (1992b) Electrogenic properties of the cloned Na$^+$/glucose cotransporter: II. A transport model under nonrapid equilibrium conditions. J Membrane Biol 125:63–79

Parsons DS, Wingate DL (1958) Fluid movement across wall of rat small intestine in vitro. Biochim Biophys Acta 30:666–667

Patlak CS, Goldstein DA, Hoffman JF (1963) The flow of solute and solvent across a two-membrane system. J Theoret Biol 5:426–442

Payne JA, Forbush III B (1995) Molecular characterization of the epithelial Na-K-Cl cotransporter isoforms. Curr Opinion Cell Biol 7:493–503

Persson B-E, Spring KR (1982) Gallbladder epithelial cell hydraulic water permeability and volume regulation. J Gen Physiol 79:481–505

Reuss L (1985) Changes in cell volume measured with an electrophysiological technique. Proc Natl Acad Sci USA 82:6014–6018

Reuss L (1991) Salt and water transport by gallbladder epithelium. In:Handbook of Physiology. American Physiological Society, Washington, Sect 6, Vol. IV, p 303–322

Reuss LA, Finn AL (1975a). Electrical properties of the cellular transepithelial pathway in Necturus gallbladder. I. Circuit analysis and steady state effects of mucosal solutions ionic substitutions. J Membrane Biol 25:115–139

Reuss LA, Finn AL (1975b) Electrical properties of the cellular transepithelial pathway in Necturus gallbladder. II. Ionic permeability of the apical cell membrane. J Membrane Biol 25:141–161

Rose RC, Schultz SG (1971). Studies on the electrical potential profile across rabbit ileum. Effects of sugars and amino acids on transmural and transmucosal electrical potential differences. J Gen Physiol 57:639–663

Rubinstein I, Segel LA (1983) Sensitivity and instability in gradient flow. In:Salanki J (ed) Physiology of Non-excitable Cells. Adv Physiol Sci 3:71–80. Pergamon Press

Sackin H (1986) Electrophysiology of salamander proximal tubule. II: Interspace NaCl concentration and solute coupled water transport. Am J Physiol 251:F334–F347

Sackin H, Boulpaep EL (1975) Models for coupling of salt and water transport. J Gen Physiol 66:671–733

Sardet C, Franchi A, Pouyssegur J (1989) Molecular cloning, primary structure and expression of the human growth factor activatable Na/H antitransporter. Cell 56:271–280

Schafer JA, Patlak CS, Andreoli TE (1975) A component of fluid absorption linked to passive ion flows in the superficial pars recta. J Gen Physiol 66:445–471

Schnermann J, Chou C-L, Ma T, Traynor T, Knepper M, Verkman AS (1998) Defective proximal tubular fluid reabsorption in transgenic aquaporin-1 null mice. Proc Natl Acad Sci USA 95:9660–9664

Schultz SG (1977) The role of paracellular pathways in isotonic fluid transport. Yale J Biol Med 50:99–113

Schultz SG, Curran PF (1968) Intestinal absorption of sodium chloride and water. In: Handbook of Physiology Sect 6, Vol. III, pp 1245–1275. American Physiological Society, Washington

Schultz SG, Curran PF (1970) Coupled transport of sodium and organic solutes. Physiol Rev 50:637–718

Schultz SG, Curran PF (1974) Sodium and chloride transport across isolated rabbit ileum. Current Topics in Membranes and Transport 5:225–281

Schultz SG, Zalusky R (1964) Ion transport in isolated rabbit ileum. I. Short-circuit current and Na fluxes. J Gen Physiol 47:567–584

Schwiebert EM, Benos DJ, Egan ME, Stutts MJ, Guggino WB (1999) CFTR is a conductance regulator as well as a chloride channel. Physiol Rev 79:S145–S166

Shachar-Hill B, Hill AE (1993) Convective fluid flow through the paracellular system of Necturus gallbladder epithelium as revealed by dextran probes. J Physiol 468:463–486

Sheerin HE, Field m (1975) Ileal HCO_3^- secretion:Relationship to Na and Cl transport and effect of theophylline. Am J Physiol 228:1065–1974

Sheppard DN, Welsh (1999) Structure and function of the CFTR chloride channel. Physiol Rev 79:S23–S45

Simon DB, Lu Y, Choate KA, Velasques H, Al-Sabban E, Praga M, Casari G, Bettinelli A, Colussi G, Rodriquez-Soriano J, McCredie D, Milford D, Sanjad S, Lifton RP (1999) Paracellin-1, a renal tight junction protein required for paracellular Mg^{2+} resorption. Science 285:103–106

Spring KR (1973a) Current-induced voltage transients in Necturus proximal tubule. J Membrane Biol 13:299–322

Spring KR (1973b) A parallel path model for Necturus proximal tubule. J Membrane Biol 13:323–352

Spring KR (1998) Routes and mechanisms of fluid transport by epithelia. Annu Rev Physiol 60:105–119

Spring KR (1999) Epithelial fluid transport – a century of investigation. NIPS 14:92–99

Spring KR, Hope A (1978) Size and shape of the lateral intercellular space in a living epithelium. Science 200:54–58

Spring KR, Hope A (1979) Fluid transport and the dimensions of cells and interspaces of living Necturus gallbladder. J Gen Physiol 73:287–305

Sten-Knudsen O, Ussing HH (1981) The flux ratio equation under nonstationary conditions. J Membrane Biol 63:233–242

Stirling CE (1972) Radioautographic localization of sodium pump sites in rabbit intestine. J Cell Biol 53:704–714

Stirling CE (1967) High-resolution radioautography of phlorizin-^{3}H in rings of hamster intestine. J Cell Biol 35:605–614

Stirling CE, Schneider AJ, Wong M, Kinter WB (1972) Quantitative radioautography of sugar transport in intestinal biopsies from normal humans and a patient glucose-galactose malabsorption. J Clin Invest 51:438–445

Strong TV, Boehm K, Collins FF (1994) Localization of cystic fibrosis transmembrane conductance regulator mRNA in the human gastrointestinal tract by in situ hybridization. J Clin Invest 93:347–354

Sullivan P, Hoff vS, Hillyard SD (2000) The effect of anion substitution on hydration behavior of the red spotted toad, Bufo punctatus: evidence for anion paradox associated with cutaneous water uptake. Chemical Senses. 25: (in press)

Sullivan SK, Field M (1991) Ion transport across mammalian small intestine. In: Handbook of Physiology Sect 6, Vol. IV pp 287–301. American Physiological Society, Washington

Suvitayavat W, Dunham PB, Haas M, Rao MC (1994) Characterization of the proteins of the intestinal Na⁺-K⁺-2Cl⁻ cotransporter. Am J Physiol 267:C375–C384

Thorens B (1993) Facilitated glucose transporters in epithelial cells. Annu Rev Physiol 55:591–608

Thurau K (1961) Renal Na-reabsorption and O_2 uptake in dogs during hypoxia and hydrochlorothiazide infusion. Proc Soc Exp Biol Med 106:714–717

Timbs MM, Spring KR (1996) Hydraulic properties of MDCK cell epithelium. J Membrane Biol 153:1–11

Tormey J McD, Diamond, JM (1967) The ultrastructural route of fluid transport in rabbit gall bladder. J Gen Physiol 50:2031–2060

Torrelli GE, Milla E, Faelli A, Costantini S (1966) Energy requirement for sodium reabsorption in the in vivo rabbit kidney. Am J Physiol 211:576–580

Trezise AEO, Buchwall B (1991) In vivo cell-specific expression of the cystic fibrosis transmembrane conductance regulator. Nature 353:434-437

Tripathi S, Boulpaep EL (1989) Mechanisms of water transport by epithelial cells. Quart J Exp Physiol 74:385–417

Tse CM, Brant SR, Walker MS (1992) Cloning and sequencing of a rabbit cDNA encoding an intestinal and kidney specific Na⁺/H⁺ exchanger isoform (NHE-3). J Biol Chem 267:9340–9346

Tse CM, Ma AI, Yang VW (1991) Molecular cloning and expression of a cDNA encoding the rabbit ileal villus cell basolateral membrane Na-H exchanger. EMBO 10:1957–1967

Turnberg LF, Bieberdorf F, Morawski S, Fordtrand J (1970) Interrelationships of chloride, bicarbonate, sodium and hydrogen transport in the human ileum. J Clin Invest 49:557–567

Ussing HH (1952) Some aspects of the application of tracers in permeability studies. Adv Enzym XIII:21–65

Ussing HH (1966) Anomalous transport of electrolytes and sucrose through the isolated frog skin induced by hypertonicity of the outside bathing solution. Ann NY Acad Sci137:543–555

Ussing H.H, Eskesen K (1989) Mechanism of isotonic water transport in glands. Acta Physiol Scand 136:443–454

Ussing HH, Johansen B (1969) Anomalous transport of sucrose and urea in toad skin. Nephron 6:317–328

Ussing HH, Lind F, Larsen EH (1996). Ion secretion and isotonic transport in frog skin glands. J Membrane Biol 152:101–110

Ussing HH, Nedergaard S (1993). Recycling of electrolytes in small intestine of toad. In: Ussing HH, Fischbarg J, Sten-Knudsen O, Larsen EH, Willumsen NJ (eds) Isotonic Transport in Leaky Epithelia. Proc Alfred Benzon Symp 34:25–34, Munksgaard, Copenhagen

Weinstein AM (1992) Sodium and chloride transport. In: Seldin, DW, Giebisch G (eds) The Kidney. Physiology and Pathophysiology, 2nd edition Vol 2. pp 1925–1973. Raven Press, New York

Weinstein AM, Stephenson JL (1981a) Coupled water transport in standing gradient models of the lateral intercellularspace. Biophys J 35:167–191

Weinstein AM, Stephenson JL (1981b) Models of coupled salt and water transport across leaky epithelia. J Membrane Biol 60:1–20

Welsh MJ, Smith PL, Fromm M, Frizzell RA (1982) Crypts are the site of intestinal fluid and electrolyte secretion. Science 218:1219–1221

White JF (1986) Modes of Cl⁻ transport across the mucosal and serosal membranes of urodele intestinal cells. J Membrane Biol 92:75–89

White JF (1989) Characteristics of chloride ion influx in *Amphiuma* small intestine. Am J Physiol. 256:G166–G177

White JF (1994) Chloride channels in epithelial cells of intestine. In: Electrogenic Cl⁻ Transporters in Biological Membranes. Eds.: Gerenscher, G.A. Advances in Comparative & Environmental Physiology 19:221–237

White JF, Armstrong W McD (1971) Effect of transported solutes on membrane potentials in bullfrog small intestine. Am J Physiol 221:194–201

White JF, Ellingson D (1989) Basolateral impalement of intestinal villus cells: electrophysiology of Cl⁻ transport. Am J Physiol 256:C1022–1032

White JF, Ellingson D, Burnup K (1984) Electrogenic Cl⁻ absorption by *Amphiuma* small intestine: Dependence on serosal Na⁺ from tracer and Cl⁻ microelectrode studies. J Membrane Biol 78:223–233

Whitlock RT, Wheeler HO (1964) coupled transport of solute and water across rabbit gallbladder epithelium. J Clin Invest 43:2249–2265

Whittembury G, De Martínez CD , Linares H, Paz–Aliaga A (1980) Solvent drag of large solutes indicates paracellular water flow in leaky epithelia. Proc R Soc Lond B 211:63–81

Whittembury G, Malnic G, Mello–Aires M, Amorena C (1988) Solvent drag of sucrose during absorption indicates paracellular water flow in the rat kidney proximal tubule. Pflügers Arch 412:541–547

Whittembury G, Paz–Aliaga A,Biondi A, Carpi-Medina P, Gonzales E, Linares H (1985). Pathways for volume flow and volume regulation in leaky epithelia. Pflügers Arch 404:S17–S22

Whittembury G, Reuss L (1992). Mechanisms of coupling of solute and solvent transport in epithelia. In: Selding DW, Giebisch G (eds) The Kidney:Physiology and Pathophysiology, 2nd edition. pp. 317–360. Raven Press, New York

Willumsen NJ, Boucher RC (1989) Shunt resistance and ion permeabilities in normal and cystic fibrosis airway epithelia. Am J Physiol 256:C1054–C1063

Windhager EE, Whittembury G, Oken DE, Schatzmann HJ, A.K. Solomon AK (1958) Single proximal tubules of the Necturus kidney. III. Dependence of H₂O movement on NaCl concentration. Am J Physiol 197:313–318

Wright EM, Mircheff AK, Hanna SD, Harms V, Van Os CH, Wallin MW, Sachs G (1979) The dark side of the intestinal epithelium:the isolation and characterization of basolateral membranes. In: Mechanisms of Intestinal Secretion. Ed.:Binder H, Liss J, New York, p 117–130

Zerahn K (1956) Oxygen consumption and active sodium transport in the isolated and short-circuited frog skin. Acta Physiol Scand 36:300–318

Zeuthen T (1977) The vertebrate gallbladder – The routes of ion transport. In:Gupta BL, Moreton RB, Oschman, JL, Wall, BJ (eds) Transport of Water and Ions in Animals. Academic Press, pp 511–537

Zeuthen T (1982) Relations between intracellular ion activities and extracellular osmolarity in *Necturus* gallbladder epithelium. J Membrane Biol 66:109–121

Zeuthen T (1983) Ion activities in the lateral intercellular spaces of gallbladder epithelium transporting at low external osmolarities. J Membrane Biol 76:113-122

Zeuthen T (2000) Molecular water pumps. Rev Physiol Biochem Pharmacol. Springer-Verlag, pp 97–151